About Island Press

Since 1984, the nonprofit organization Island Press has been stimulating, shaping, and communicating ideas that are essential for solving environmental problems worldwide. With more than 1,000 titles in print and some 30 new releases each year, we are the nation's leading publisher on environmental issues. We identify innovative thinkers and emerging trends in the environmental field. We work with world-renowned experts and authors to develop cross-disciplinary solutions to environmental challenges.

Island Press designs and executes educational campaigns, in conjunction with our authors, to communicate their critical messages in print, in person, and online using the latest technologies, innovative programs, and the media. Our goal is to reach targeted audiences—scientists, policy makers, environmental advocates, urban planners, the media, and concerned citizens—with information that can be used to create the framework for long-term ecological health and human well-being.

Island Press gratefully acknowledges major support from The Bobolink Foundation, The Curtis and Edith Munson Foundation, The Forrest C. and Frances H. Lattner Foundation, The Freedom Together Foundation, The Kresge Foundation, The Summit Charitable Foundation, Inc., and many other generous organizations and individuals.

The opinions expressed in this book are those of the author(s) and do not necessarily reflect the views of our supporters.

The Light Between Apple Trees

Also by Priyanka Kumar

Conversations with Birds
Take Wing and Fly Here
The Song of the Little Road

The Light Between Apple Trees

Rediscovering the Wild Through a
Beloved American Fruit

Priyanka Kumar

 ISLANDPRESS | Washington | Covelo

© 2025 Priyanka Kumar

All rights reserved under International and Pan-American Copyright Conventions. No part of this book may be reproduced in any form or by any means without permission in writing from the publisher: Island Press, 2000 M Street, NW, Suite 480-B, Washington, DC 20036-3319.

Library of Congress Control Number: 2025936288

All Island Press books are printed on environmentally responsible materials.

Manufactured in the United States of America
10 9 8 7 6 5 4 3 2 1

Art in chapters 11, 12, 16, and 19 in the public domain. All other photographs taken by the author or courtesy of the author.

Keywords: America, American Southwest, apple, biodiversity, cider, crabapple, ecosystem, foraging, Golden Delicious, Granny Smith, Henry David Thoreau, Himalayas, Kazakhstan, land, nature, New Mexico, Newtown Pippin, orchard, Red Delicious, Thomas Jefferson, Tian Shan, Winesap

For Michael

I broke with the sun and stars.
I let the world go.
I went far and deep with the knapsack of things I know.
I made the journey, bought the useless, found the indefinite,
And my heart is the same as it was: a sky and a desert.

   —Fernando Pessoa

Contents

Preface — xiii

AUGUST
  1. An Apple for Ursula — 1
  2. In Search of Feral Apples — 9

SEPTEMBER
  3. The Venerable Crab — 17

OCTOBER
  4. Seeking Celestial Apples — 27
  5. The Fruitery — 37
  6. Ten Thousand Varieties — 49
  7. The Flow of Energy — 59
  8. Pulling the Crank — 67

NOVEMBER
  9. Wild Horses — 81
  10. Industrial Fruit — 93

DECEMBER AND JANUARY
  11. The Archbishop's Garden — 105

FEBRUARY
  12. Community Making — 115

MARCH
  13. Mapping the Orchard — 123

APRIL
  14. The Generosity of Blossoms — 135

MAY
   15. Cow Creek Runs Through Paradise   147
JUNE
   16. True Wild   159
JULY
   17. Love Notes from Nature   165
AUGUST
   18. Summer Lake   179
   19. The Hidden Rose   187
SEPTEMBER
   20. The Songs of Ancient Trees   195
OCTOBER
   21. Newtown   207
   Epilogue   221

Acknowledgments   225
Select Sources   229
About the Author   240

Apple blossoms. Photo by Priyanka Kumar.

Preface

The spark of this narrative is a day spent in a Himalayan apple orchard when I was maybe five. Every autumn, my mind returns to that day, but I cannot crack open the experience or figure out why I keep circling back. I can scent and taste the childhood memory, but the green images slip away like phantoms. At last, I felt stirred to journey through old orchards, and I wondered if the practice might help unfurl the memory and its hidden meaning.

The foothills of the Himalayas were the sanctuary of my childhood. I spent my first decade exploring their resplendent gifts, and I felt adrift when my family moved away. Since then, I have struggled to make a home in places where I seemingly didn't belong. In time, I would come to see myself as a citizen of the earth. Each portion of our planet holds some sliver of nature, maybe a yellow-breasted chat caroling from an apple tree, each an invitation to peer more closely. Chasing the childhood memory was never about locating the Himalayan orchard that had enthralled my five-year-old self. I was searching instead for what the orchard had evoked in me. As Joseph Campbell might say, I was "caught," even ensnared, by apples—especially wild and feral ones. I saw orchards differently: I wondered if an orchard could become a gateway to wilderness.

My odyssey deepened into a search for forested orchards. I was attracted to mature orchards as havens, providing habitat and foraging for birds and keystone mammals such as bears, as well as offering sustenance to us. The scientist E. O. Wilson likened an ecosystem to a web woven by diverse plant and animal species who cooperate and interact in complex ways in a singular geographic area. "This is the assembly of life that took a billion years to evolve," he wrote, hailing biodiversity as the key to the maintenance of the world as we know it. So, when species diversity and gene flows grow impoverished, ecosystems cannot remain resilient or support the "assembly of life" that depends on them. The story of apples mirrors that loss.

Ecosystem problems can feel remote when we don't have meaningful access to wilderness areas. As a naturalist, I am all too aware of the barren surroundings—metal, glass, and plastic—in many of the places where I find myself. Swamped by industrial habitats, it is challenging to *see* the silent parade of one million species facing extinction. So, I sought out micro-wilderness areas at the boundary of the urban and the wild. Could these miniature wilds enable us to see better? Here, I was surprised to discover feral apple trees, also called seedlings or pippins, and an array of wildlife from the hummingbird to the bear. Could these in-between spaces yield insights that might untangle our too-linear relationship with nature? Could the magic of feral groves dismantle the false dichotomy between civilization and nature?

In tiny and grand ways, many of us have glimpsed what is sacred in nature. Still, we view nature as belonging to another realm, separate from us. Bringing green sacredness into our everyday lives could break the suffocating walls we've cemented between ourselves and nature over two industrial centuries. While my childhood memory nudged me toward apple trees, the American Southwest, where I now live, is also a source of unexpected history beyond the familiar settler narrative, shining a fresh light on America's apple story; apple trees have thrived here for a few hundred years and are still woven into the lives of locals and wildlife. To absorb the significance of biodiversity loss and climate change, it helps to engage with the biosphere that enables life on this planet, including our own. This book is one map of what such engagement could look like.

As my love for apple trees deepened, my children's reaction amused me: They have sharp opinions about how many hikes we can tuck into a single week, but they followed me spiritedly on fruity paths. As a child, I was a devotee of green—meadows, trees, mountains. Trees were mysterious beings who animated my days and allowed me to walk through in peace and attune my body to nature's music. These pages tell the story of how the humble and iconic apple tree can lead us back to nature and see biodiversity's vital role at a time when we are grappling with too many catastrophes. Trusting a childhood memory, I stepped into the world of apples and found myself going deeper into the forest.

Chapter One

An Apple for Ursula

AUGUST

An apple tree near the River Trail. Photo by Priyanka Kumar.

I

My feet quicken as I hike to an apple grove near the Santa Fe River. Walking in the footprints of a black bear who strolled through here last night, I maneuver around copious bear scat, dotted with berries. The trail meanders under a canopy of spindly elms and maternal cottonwoods, and the riverbank is fringed with willows. Creatures skitter away in my path: Skinks with turquoise tails glide over rocks; a bull snake undulates across the trail; water striders dance over the glassy water; and a cricket hovers in the air like a hummingbird. The river belches a grumble; we are in a megadrought, and the water level is perilously low. When I reach a clearing, the late August morning is lustrous and the air shimmers with light. The insistent music of a black-capped chickadee fades away, but my heart thrums as I approach the spot that struck me with the force of a discovery the previous week.

Tall, wild grasses are flattened into a wide circle about the right size for a mature *Ursus americanus*, black bear, to sleep in. The flattened area is tawny and dry, but the surrounding grass retains the bleached green of summer. The size of the circle alone wouldn't be enough to go on if it weren't for the fur in the dirt. Bits of bear fur are scattered on the ground, dark brown with fine white strands. I let out a long breath—I can all but see Ursula the bear curled up and sleeping here in the dark. A fortunate bear who has an orchard for her refuge!

The apple trees in this grove were planted several decades back, and they grace a patch of private land like mellifluous pillars. I walk on to a tall apple tree favored by Ursula, who has marked the circumference around the trunk with her scat. The tree bark is smooth and metallic gray-brown, and the colors shift with the day's light, conjuring Rilke's lines in the *Book of Hours* in which he speaks directly to God: "I want to utter you. I want to portray you not with lapis or gold, but with colors made of apple bark."

Sidestepping Ursula's scat, I gently pull down one of the tree's branches laden with small russet apples. A worm has bored through one apple and I let it fall to the ground—for Ursula. Many of the apples are strung on high branches, and I imagine the bear up on her hind legs stretching to reach them. Bears and apple trees likely evolved together, like a symphony perfecting itself. In the late Pleistocene, before megafaunal mammals became largely extinct, they played a role in dispersing large fruits in ecosystems around the earth. Archaeobotanist Robert Spengler points out that the apple's small seeds allow the fruit to be more easily dispersed by midsize omnivores and bears, or megafaunal mammals (living or extinct animals who are larger than humans or at least heavier than ninety pounds). Apples played their part to strengthen the mutualism: Early apples or June apples are aromatic, bright, short-lived, and sometimes watery tasting, but they probably evolved to attract young bears, who can more easily climb into a tree's lower boughs. In a separate movement of the evolutionary symphony, the claws of bear species evolved and made it easier to catch fish *and* acted as rakes to harvest fruit.

An adult black bear can stand between five and six feet tall and couldn't reach much of the fruit I am looking at—which is why Ursula sometimes snaps tree branches. An acquaintance, Tim, lives just uphill from this trail. He woke up on a recent morning to find his wire fence partially flattened and two branches of his pear tree broken—the work of a hungry bear.

"I got off easy," Tim told me with a dry smile. He lost several pears but found it exciting to think about the bear wandering through his property. I sense that Tim is a kindred spirit. Ursula, a female bear, could weigh anywhere between ninety and three hundred pounds. So, Tim is understandably stirred that a megafaunal mammal sometimes visits his yard.

In the orchard, I bite into a small red apple—sweet with hints of tartness, like the early fall morning. No wonder Ursula favors this tree. Moments later, phenolic compounds in the apple react with oxygen in the air, and the exposed flesh turns a rust color. This apple has a high level of polyphenol oxidase activity and may be unusually rich in iron. Compared to commercial apples, the ones I forage are likely richer in nutrients since they grow in old orchards, which tend to have biodiverse soil.

During the harvest season, I studiously avoid the supermarket apple.

One fall, however, my husband Michael got some organic Golden Delicious apples from British Columbia at the Natural Grocers in our hometown, Santa Fe. They turned out to be cardboard-like and so hollow that if you pinched the skin, it gave under your thumb. We composted the whole lot. The legendary scientist Aimak Dzhangaliev, who studied the genetic diversity of Kazakhstan's wild apples, once said: "A widespread cultivar like the Golden Delicious is simply not that good. It does not have good nutritional value nor much disease resistance.... We not only have to do better plant genetics work on these apples, but we have to fight against [reiterating] the errors of history." But what are these *errors* of history? Having observed the rhythms of life in forested orchards over several years, I suspect that crowded monoculture orchards are one key error that led us to the corrugated cardboard–textured apple.

I have dived headlong into foraging fruit. I began my foraging career at the age of five, when I wafted like a butterfly through apple trees in the foothills of the Himalayas, but it took the pandemic for my lifelong passion for apples to ripen. During the pandemic years, I discovered more fruit trees than one might expect to find in the high deserts of the American Southwest and the Pacific Northwest—apricots in July, peaches and plums in August, a variety of apples in September and October. As my foraging deepened, I grew aware that fruit trees were connecting me to my past. The act of picking apples is in me; this is how I lived as a child.

In my Himalayan days, I was unselfconscious—a butterfly scarcely thinks of who she is—as I wandered freely in one of the planet's leading biodiversity hot spots. The wildness of this region in Himachal Pradesh has left a deep imprint—I still dream about the state animal, the snow leopard, and chase the state fruit. My recent plunge into fruit trees led me to wonder why my first visit to a Himalayan orchard had been so meaningful. Was it because my father had initiated me into the universe of fruit trees? I lost him to political violence when he was fifty-four—and I was a graduate student. The loss stays in me like a primal ache, but I come closest to transcending it when I engage in the practice of foraging. Picking fruit from trees draws me closer to the elemental life we shared. After wandering through orchards and the adjacent micro-wild, a wave of aliveness surges in me.

At a time when our community had atrophied to Zoom meetings and real contact with extended family was cut off by the temporary closure of the US-Canada border, my family truly became an island unto ourselves. Yet it wasn't loneliness that drove me to apple trees, but a search for riches, including solitude. I searched for life's essence, which has a different fragrance than the everyday path we hazily walk on. The universe is nearly at our doorstep, and we scarcely know it. I sought the exhilaration that comes from being fully present in the wild. An ancient Indian concept illuminates how the emotional flavors we embrace can set the mood or tenor of our lives. Rasa means juice in Sanskrit and suggests the vital essence of an experience, person, or work of art. Seeking fruit straight from trees revitalized me and infused my life with a strain of rasa.

How did apple trees offer sustenance and become an anchor at a time when so many of us were feeling unmoored? The answer may have to do with how my apple journeys were taken in empathy with the wild. The wild brings us primal discomforts but also forges kinship with trees and bears and thwacks us to life with heart-stopping surprises. I couldn't have anticipated that my search for wild apple trees and historic orchards would become an exploration not only of biodiversity (how feral apple trees support bears like Ursula) but also of America's love affair with apples. While foraging, I reap the fruit of my father's land ethic, the apple seeds the Spanish and the settlers brought to America in the 1600s, the seeds John Chapman grew into saplings and often gave away, and the young saplings that the French missionary Jean-Baptiste Lamy and others grew into mature trees in the 1800s—scion wood from those trees would later be grafted onto newer rootstock, and the process repeated until were born the trees from which I pick apples today. Americans once lived in real kinship with apples, but that relationship has frayed in modern times. I grew aware of the fragile status of the orchards where I foraged, and it felt urgent to understand the story of apples—how they had evolved, were domesticated, and became beloved—and whether the past could offer insights to prevent us from reiterating the errors of recent apple history.

II

A primal flowering rose evolved on earth under temperate climate conditions some 80 million years ago. How long ago was that? Scientists in China recently discovered the smallest known non-avian dinosaur egg, the size of a small strawberry, which was dated as being 80 million years old. Over thousands of millennia, fruit trees, including apple, cherry, peach, pear, and plum, all evolved from that early rose—and belong to the Rosaceae family. Conveniently, bees already existed on earth before the arrival of flowering plants. Apples are believed to have originated as a cross between an ancient plum and the meadowsweet plant in the genus *Spiraea*, in an area between the Caspian and the Black Seas.

That being said, the origin of fruit trees is one of the biggest unanswered questions in domestication studies. Only recently have genetic studies yielded tantalizing clues about how the wild apple, *Malus sieversii* (named after botanist Johann Sievers, who first described it in the late 1700s), was domesticated. During the Eocene, 56 to 34 million years ago, the Rosaceae family diversified and forged a stronger mutualism with animals, including birds; small fleshy fruit grew prominent, attracting avian seed dispersers. The oldest bee fossil that can unambiguously be placed in an existing family of bees also dates from this period.

Spengler, the archaeobotanist, suggests that in the late Miocene, 23 to 5 million years ago, apples grew even larger, and megafaunal mammals began to eat them and disperse seeds. During the late Pliocene, some 2.6 million years back, vast temperate forests flourished in central Asia. Scientists have speculated that the ancestors of fruit-eating birds such as the azure-winged magpie flew through these forests, which ranged from western Europe to eastern Asia in the late Tertiary and early Quaternary. These ancestral birds would have gorged on a wilder version of a bird-dispersed apple we know today as *Malus baccata*. On the River Trail where I seek apples, I have run-ins with black-billed magpies, building or defending a nest or chattering away with spirited river or rock. Over millennia, apples have seduced all creatures, from magpies to bears. And humans.

In time, ancient temperate forests graded into foothills, steppe, and desert. One key region, the Tian Shan, now in Kazakhstan, remained temperate and became girdled by harsh deserts. Discrete boundaries are considered to be a critical ingredient in the evolution of species—so it was that propitious conditions arose for wild apples to perfect themselves in the Tian Shan, the Celestial Mountains. Unlike some fruit, apple trees are less likely to grow from rotting apples fallen on the ground; so, apples need allies such as birds and mammals to disperse their seeds. In addition, local humans played a role in dispersing apples. At a village site in the Tian Shan in Kazakhstan, archaeologists discovered an apple seed that was dated to the end of the first millennium BCE. In the twentieth century, in this very area, scientists would discover apple forests growing wild in the Tian Shan range.

In America, when I seek wild apple trees, I use the term "wild" in a cultural rather than a scientific sense—in the way that Henry David Thoreau called those apple trees wild that were self-seeded or ungrafted and irregularly planted. Biting into such feral apples is intoxicating. Standing under a tree, enveloped in crisp bluish air, when I taste an apple, all I'm doing is tasting an apple. While savoring a foraged apple, I become aware of the grass under my feet and the reddish globes hanging enticingly from trees of wildly diverse shapes and sizes. A flicker might fly past and a chickadee sing. This is enough; this is rasa, a hidden spring of joy. Having experienced dizzying changes in life, chiefly migration and mortality, I yearn for this at least to stay the same. I grow exultant at the thought of discovering "wild apples"—and recall how Thoreau had raved about them.

When Thoreau trekked beyond his neighborhood on longer hikes, he sometimes turned to wild apples (*agrestia poma*, as he called them) for refreshment. He relished the tart fruit while admitting that it tasted good only in fresh air. In the mid-1800s, Thoreau saw that wild apple trees were already on their way out. "The rows of grafted fruit will never tempt me to wander amid them like these," he wrote. "But I now, alas, speak rather from memory than from any recent experience, such ravages have been made!"

In the industrial age, land in Concord, across America, and around the world became highly monetized, a process that has speeded up a thousandfold in our day. Since 1970, the human population has doubled whereas wildlife populations have declined by roughly 70 percent. Scientists have zeroed in on habitat loss as the primary driver of biodiversity loss—that includes the loss of thousands of apple varieties. We have erected countless structures where the glories of the living earth once lay. It is no surprise that we are surrounded by what can only be called scraps of ecosystems. It grows ever harder to chance upon wild fruit trees, but it is an invigorating and essential pursuit—like knocking on the door of the universe. Wondering if the tree that Ursula favors had a wild cousin nearby, a self-seeded tree, I began my search.

Chapter Two

In Search of Feral Apples

AUGUST

The River Trail. Photo by Priyanka Kumar.

I

I have been hiking the River Trail since early in the pandemic, when Michael and I started bringing our two children here. Pika, then four and a half, showed us the trail, proving that children can be our best teachers. We had stopped by to take a last look at the tree-rich playground of Pika's preschool, which had recently shut down along with everything else, when she literally tugged at our sleeves to take us to "the bridge." She led us to a trail that her class sometimes took to the river, where they would sit on the riverbank to eat lunch. Pika showed us a single-log bridge, which we gleefully crossed, then Michael and I kept walking, trailed by Pika and her older sister, Mia. The trail turned out to be longer than Pika had imagined. When at last she hotly declared that she couldn't walk anymore—"My legs are tired!"—Michael ended up carrying her back. In the coming weeks, in this rare riparian habitat, we found isolated stands of apricot and apple trees. I had encountered a scattered wealth of mature apricot trees elsewhere in the city, and a sprinkling of plum and peach trees in unexpected nooks, sidewalks, and gardens. Sometimes I would stumble upon a lone surviving diehard from what may have once been a thriving grove. But a development spree is causing habitat to dwindle even for the fruit trees that have endured.

I feel a surge of desire to protect historic fruit trees, for they are a living connection to history and science, and the wild. Since my childhood, I have sought food that is close to the soil. When I couldn't pick fruit or herbs, I ate flowers. My body sensed that food taken from nature's arms nourishes deeply, though it can also sicken us if we are foolish and ignorant. I marvel at land that supports us even after we've thrown it out of balance. In the forests of Assam, in the eastern foothills of the Himalayas, I lived in the land of wild bananas, or wild *kol*, where unusual varieties such as the hairy or pink banana and the petite *kaskol* were also grown. Indigenous communities relished feasts on banana leaves and used the flower and root of this herbaceous plant, in the genus *Musa*, in traditional dishes. As a child, I saw these Indigenous peoples as being highly intelligent—for they preferred eating outdoors, as I did, and afterward we could easily compost our leafy plates. In our Haflong bungalow we had a fruit storage room where I routinely cradled a branch laden with a hundred petite bananas. This narrow room with its sulfurous odor (from the powder sprinkled to ripen the green bananas) was my favorite place to read and to think about snakes and other wild animals I'd encountered. The narrow and long fruit storage room was the first space I had to myself—it was my first study. Strangely, the room presaged where I find myself now: a naturalist intrigued by the place of fruit in nature and in our lives. Perhaps this is another reason why a hidden spring of rasa began to flow when I discovered the covert stands of fruit trees in Santa Fe.

Walking among the fruit trees of the River Trail, the carefree Assamese days returned to fortify my butterfly self. I wanted to realign myself with the land and its animals and embrace a way of being that resists the fragmented lives we live, sundered from nature. It is true that I was walking in the scrap of a riparian ecosystem when my heart lay in exploring relatively intact ecosystems. But wilderness areas aren't always within reach. The forests around us are battered by drought and wildfires, and they can no longer absorb a quarter of the carbon emissions from the atmosphere each year as they once did. I would have to make do with what was at hand.

Land literacy can fuel a radical life, and I was soon writing to the city's mayor about conserving our micro-wild and ecosystem scraps. I got back blandly polite responses, however, and felt bewildered. Meanwhile, more land—pristine swaths of desert—was being devoured, sometimes when the state auctioned off treasures in its trust to the highest-bidding developer. Glorious small orchards have similarly been abandoned or decimated across the country. Even relatively quiet communities have experienced development spikes. It seems that no amount of land devoured is finally enough.

While mentally composing yet another letter to the city council, I found myself growing dispirited, and I paused. First, I needed to deepen my practice of connecting to the wild and answer my questions about the origins of the apple trees I saw. I also wanted to experience more fully the fruit trees that remained so that I could articulate why they should be valued.

Is it worthwhile to experience a scrap of an ecosystem? I would try and find out. My own neighborhood, where I walked daily, was experiencing a serious loss of biodiversity; some cues were an uncanny silence of plant life and a relentless ebb of insect and bird song. Birds I had counted on for over a decade had stopped coming to the area. I gazed at each clay-colored hill that hadn't been razed flat (most have been). When I stumbled upon a grove of trees left intact, it felt like a discovery. "This grove must be where the bushtits roost," I told myself. "That hill must be a refuge for the coyotes, or perhaps a bobcat." One evening, I nearly stepped on a juvenile snake and a nighthawk flew over me, encouraging my faith in the micro-wild. Such remnants of the wild are rare, however, and feel almost sacred.

When I experienced that wild scraps could move my spirit, trailing the river on weekends became an essential part of my practice. I couldn't have anticipated that the months from July to October would come to mean a luscious and tart procession of fruits, beginning with apricots and peaches in mid-summer and climaxing in bluish-purple plums and dazzling apples. While picking apples, I follow Thoreau's dictum: "Surely the apple is the noblest of fruits. Let the most beautiful or the swiftest have it. That should be the 'going' price of apples."

II

All children stomp and whine about hiking, but the gurgling river helped entice ours, and the paths that led to the fruit trees became the yellow brick roads they grew to savor. While Ursula the bear roams this trail when humans are out of sight and a deer skeleton signals a mountain lion's hunger, the animals we've befriended here range from line and bull snakes to downy and hairy woodpeckers in episcopal caps, pecking hypnotically at tree barks. The fruit we forage is eaten on the spot or ripened on a windowsill. The apples tend to be wormy around the core and the clean parts must be sliced off. We stew the slices and sprinkle in cinnamon, cardamom, and cloves, for a healing Ayurvedic breakfast.

The primal pleasures of picking feral apples may be forgotten today, but the evangelist of apple love remains a staple of our classrooms. Pika would learn in first grade about Johnny Appleseed and the tall tales that Americans tell about him: No, Johnny's skin wasn't so thick that a snake couldn't bite through, nor did he play with bear cubs. It's more likely that Appleseed, or John Chapman, was vegetarian and went out of his way not to harm animals. In a coffee-sack shirt and wide-brim felt hat, he tended to wander away from his crowded childhood home until, finally, in late 1805 or early 1806, at the age of thirty, he stuffed apple seeds into leather pouches and steered his canoe upstream on the Muskingum and Licking Rivers.

While picking apples near the Santa Fe River, I experience a fellow feeling with Ursula: I follow her tracks and, the next day, she'll follow mine. Bears and apple trees are braided together, and Ursula is a lynchpin of the riparian community I am observing. My practice is to let the material world fade and instead tune in to nature's throbbing connections. As a megadrought rages in the Southwest, acorns are expected to wane in abundance and apples will become more vital for bears, who must fatten up before going into torpor or deep sleep at the end of October. Leaving a few apples under a tree for Ursula is my way of respecting a fellow sentient.

In my neighborhood, a woman drove straight into a female bear and killed her. Two cubs were accompanying their mother. One ran away, but a couple of hours later, Animal Services located the cub squatting on a low tree branch; the second cub's whereabouts are unknown. The captured cub was taken to a wildlife center with the intention of releasing him in a forest in January. I don't know how or why this woman drove her car into a bear; presumably, it was an accident. The act may have sealed the fate of the remaining cub, however. The New Mexico Department of Game and Fish cautions that killing a mother with cubs often means that the cubs will die too.

I don't mean to romanticize Ursula, nor am I eager to once again hear a bear's growl or her otherworldly snorting. After an encounter with five bears several years back in Northern California, I learned to be wary. I understood then that backing away from a bear is a far better strategy than running away and that bears are considered to be a threat only when they are surprised or protecting their cubs. Years later, now that I have children, I find myself appreciating the connection we share with bears: Fueled by protective instincts, mother bears stay with their cubs and raise them for relatively long durations—from eighteen to thirty months.

III

As a forager I keep a lookout for wild fruit, but I also observe what fruit other animals—big and small—eat. In August, I stayed for a month in a century-old cabin in northern New Mexico, which forester Aldo Leopold designed and lived in with his young wife, Estella. In the mornings, I would climb up the rocks behind the cabin to an immense boulder from where I could peer into a ponderosa pine glade. Early one morning, I saw a chipmunk approach a prickly pear cactus and turn away on seeing no fruit on it. The week before, five-year-old Pika had climbed a boulder near the cabin and seen fruit on a prickly pear.

The next day, the reddish-yellow fruit had already rolled down the massive boulder.

The day after, the fruit intriguingly vanished.

Pika asked about it. Now I could tell her that a chipmunk had accepted the gift of the cactus and consumed the fruit, seeds and all. The chipmunk doesn't waste. At dawn, chestnut fur glowing, the chipmunk would devour the feathery-pink flowers of the native Apache plume. Black-and-white stripes along the face and back set off a delicate russet beauty. The chipmunk consumed the pink, flowery strands while the rose strands in the sky resolved into a glaucous daybreak. In the Gambel oak thickets that skirted the cabin, chipmunks relished the yellow-orange cushiony gall that grows on the pale underside of oak leaves. From atop the trees, they let out terrifically loud *psstt-psstt* calls and, on a boulder, a chipmunk tail flicked in beat to the call.

Like chipmunks, Ursula favors wild fruit, especially apples and berries. Berry seeds pass through her gut unbroken and, on the ground, her scat acts as a fertilizer in which the seeds can germinate. Ursula eats far more berries than birds do, and disperses significantly more viable seeds. So, she plays an extraordinary role in revitalizing plant communities. A single pile of bear scat planted in a Rocky Mountain National Park greenhouse yielded twelve hundred berry bush seedlings—one-third were chokecherry, a difficult plant to grow from seed, and the rest were Oregon grape. Scientists and researchers such as Taal Levi at Oregon State University acknowledge that bears are like farmers—by planting seeds everywhere, they promote a vegetation community that goes on to feed them. It follows that a decline in bear density adversely impacts plant and animal communities that rely on wild fruit. When humans disrupt nature's finely tuned balance by fragmenting more land or taking out an errant bear who has eaten *our apples*, the disturbance reverberates throughout the forest.

The grove I hike to near the Santa Fe River has over a dozen apple trees. We return to this place almost weekly. With development pressures picking up, I hear and see two large houses go up as I walk the beloved trail. This riparian zone, an ecological strip, is suffering more rips; at what point will the micro-wild crumple? The question nags me because the vast majority of riparian ecosystems—biodiversity heroes—have collapsed across the country. Just as apple varieties such as the dappled yellow-red Champion have become functionally extinct—they are no longer known or grown in meaningful numbers and according to the Montezuma Orchard Restoration Project, "their unique characteristics are at risk of being lost forever"—species who rely on at-risk ecosystems, such as *Empidonax truillii*, willow flycatcher, have also become threatened or endangered with the precipitous loss of streamside habitat. The National Park Service estimates that riparian zones in the Southwestern United States make up less than 2 percent of the land area but support the highest density and abundance of plants and animals compared to any other habitat type.

I divert my eye from the construction by taking note of the skip in Pika's step. With each hike, she walks farther, and we rarely carry her now. She calls the flattened circle with Ursula's fur "the place where the bear lives." Studying the area one morning, I noted that a little trail connects Ursula's circle to the apple tree she favors. After exploring this grove, we walk along the river until "No Trespassing" signs force us to turn around.

One fall morning, half a mile from the apple grove, I unexpectedly came upon a lone apple tree. I paused, surprised: In front of me grew a tree that a bear might well have planted. Tucked into a corner of the trail, flanked by a willow and an elm, the tree I had stumbled upon was almost hidden but for its telltale fruit. It grew high, for there was little space to grow wide. Most of its apples were unreachable, but I wedged myself into the thick vegetation to see if I might be able to grasp a branch. I returned to the trail with some honest scrapes and a handful of apples. The apples were smaller and redder, with deeper flavor than the apples from the grove. Surprisingly, none of these apples had worms. Maybe codling moths hadn't found this loner or else the tree had exceptional resilience.

Thoreau's regard for wild apple trees shimmers when he writes: "Here on this rugged and woody hill-side has grown an apple-tree, not planted by man, no relic of a former orchard, but a natural growth, like the pines and oaks."

Offering this fruit of "a natural growth" to Pika, and biting into one myself, I decided that these sweetly tart apples were fine refreshment on a sunbaked hike. She agreed, delighted to rest with a fruit she hadn't expected in this part of the trail. I felt strangely buoyant. I had found a wild cousin of Ursula's tree. What was more exhilarating was the discovery that life stirs in this fragmented riparian strip in ways that mirror an actual riparian ecosystem. The taste of wild apples had hope mingled in it.

Chapter Three

The Venerable Crab

SEPTEMBER

Pockets swollen with apples. Photo by Priyanka Kumar.

I

An unseasonably warm September has grazed the state's record maximum temperatures for the month. I wake up as the day's light spills like milk into the indigo sky. The temperatures don't spell autumn, but I keep a lookout for jewellike migrants. A small yellow bird with beady black eyes, and a shapely black cap on sunny head, flits among the junipers. *Cardellina pusilla*, the Wilson's warbler, pauses in my garden each year on his way south to Mexico or Costa Rica. This shy warbler, yellow-bodied and olive-backed, breeds in New Mexico and winters across the coasts of Mexico and Central America and from the Caribbean lowlands to the cloud forests of the continental divide. Seeing the warbler searching for insects and berries in my garden all but transports me to a cloud forest. The presence of migrants signals that autumn is here, but it has been terribly warm and the hummingbirds haven't left—they go on sipping from the Salvia "raspberry delight" in the rose bed. Aware that warming has disrupted the migration patterns of monarch butterflies, I find myself talking to the hummingbirds, urging them to leave before it's too late.

Walking through the garden, I note with relief that the morning air is tinged with coolness. This is my sylvan hour, before the morning exhales hurry and heat. *Malus ioensis*, a prairie crabapple tree, was once the centerpiece of the garden, but it didn't sprout a single leaf this spring. Yet the tree comes to life when *Thryomanes bewickii*, the Bewick's wren, screeches with a tart musicality; the short, shrill calls serve as my summons to be present. During the sylvan hour, the world can and must wait. Standing at the edge of a world that is shackled with industrial and digital busyness, I devote the early morning to avian and arboreal friends. The wren shuffles along the ground in search of beetles as I walk over to the crabapple. A favored perch of many birds, the prairie crabapple is native to the central US. The tree's near-demise depresses me, and I shift my gaze to the walking onions. They are sprouting from bulbs I foraged and put into the ground a little late this year. Planted near the base of a young plum tree, the onions drink from the water we give the plum, and their scent can deter deer. Such symbiotic connections are what I seek to restore among the plants. Long ago, I dreamed about making a garden that would be quivering and humming, buzzing, shrilling, and flowering.

What hurts is that the prairie crab has stopped being alive in any of those ways.

The name crabapple originates from the Nordic word *skrabba*, "fruit of the wild apple tree," and also the Celtic word *crab*, which means sour; in old English, it translates as "bitter" or "sharp-tasting." Sour or not, children pick fruit so long as they can reach it. Later that morning, an hour north at the Los Luceros Historic Site, while watching Mia gather peach-colored crabapples, I accidentally step on a path that *Pogonomyrmex barbatus*, harvester ants, take to their anthill. A local couple standing ten feet away warn me about the ants. I step back, but large red ants are already swarming up my foot. Shaking them off, I ask the couple, "What do people do with crabapples?"

"The only thing I've heard of is they make jelly from it," the man says.

I nod. European settlers routinely made preserves and vinegars with this small, fleshy fruit while Native Americans stored it underground until spring to let it sweeten. Generous trees, crabapples make rich clustering blooms, pinkish-white clouds that attract a bevy of pollinators. Those bees and wasps then go on to nearby apple trees, which cannot self-pollinate and benefit immensely from proximity to crabapple trees. My prairie crabapple's crown once served as resting and foraging grounds for an army of bees, edgy black-chinned hummingbirds, squeaking goldfinches, curve-billed thrashers, chattering evening grosbeaks, and even a stately roadrunner. Over the last fourteen years, the venerable crab has gone from erupting into a magnificent fuchsia canopy each April—I could set a calendar by how the tree bursts into bloom in mid-April—to producing a lone flower last spring. It is yet another victim of our prolonged drought. The forlorn birds are all but begging me for another crabapple or two. I also find myself dreaming about getting some varieties of cultivar apples to go with the new crabapples. Planting a diverse variety of fruit trees would increase the odds that some will survive the vagaries of a changing climate.

II

Crabapples have turned out to be more important in the origin story of apples than almost anyone could have anticipated. My prairie crabapple is one of four crabapple species that are native to the United States. It's the wild European crabapple, *Malus sylvestris*, however, that is generating buzz in scientific circles. Sylvan means "of the woods," and I think of *Malus sylvestris* as the woodland crabapple, which prefers the "wet edge of the forest" and began to expand across Europe at the end of the last ice age. It is believed that Europeans relished crabapples long before they tasted a cultivar apple. A large number of wild crabapple remains have been found in European archeological sites dating back to the Neolithic era (7500–4500 BCE).

Elsewhere, though, in the Jordan Valley and Anatolia, locals were eating apples around 6500 BCE. A significant archaeobotanical discovery all but holds us by the hand and walks us back to antiquity: Remains of some kind of apples were found in the royal tomb of Queen Pu-abi near the ancient city of Ur in Mesopotamia, dating to the late third millennium BCE. Found on a saucer, "one of the most intriguing finds were the perforated halves of small dried apples," scientists explain. "It appears that they were cut transversely in half when fresh and threaded on a string and dried before being deposited in the tomb." The fruit was "very small and wrinkled in appearance." Because it is too arid and warm in Mesopotamia for crabapples to grow wild, scientists came to believe that these apples were likely cultivated. The discovery solved a mystery from the same time period concerning cuneiform writings that describe small fruit on strings: "The sometimes disputed translation of 'haihuru'" is not apricot, but apple.

Mesopotamians seem to have been apple lovers, and some scholars argue that grafting—a key tool of apple propagation—was invented either there or in ancient Persia. A significant clue is a fragment of a Sumerian cuneiform tablet, and a theory about how grafting technology may have developed. The tablet, from roughly 1800 BCE, describes the movement and replanting of grapevine shoots. Barrie Juniper and David Mabberley interpret this record to refer to grafting, though they acknowledge the difficulty of Sumerian translations and the tentativeness of their interpretation. Their fascinating theory suggests that grafting originated in Mesopotamia as a response to a gradual salinization of the soils and consequential decline in their grape yields. Farmers would have selected for grape rootstock that is tolerant of salty soils and grafted salt-intolerant grape scions onto hardy rootstock. However, other scholars contend that this tablet could be referring to the vegetative propagation of grapes through cuttings, and the original translation of the Sumerian script into English makes no mention of grafting.

Apples of some size were certainly present in Mesopotamia by the time that Alexander the Great traveled near Babylon in 323 BCE. His records describe him ordering his boat crews to practice their battle readiness by pelting each other with apples! One can only speculate on what kind of apples these may have been, how they got there, and whether grafting was involved. Where and when grafting was invented is at best speculative, though most scholars agree that it was outside of the Mediterranean region. Grafting may also have had an independent origin in China, around 2000 BCE, through the development of silk and the dependency of silkworms on white and black mulberry leaves.

From roughly 130 BCE to 1450 CE, the Silk Route vitally connected China in the east to central Asia before radiating west to Europe. *Malus domestica* (think of it as the domestic apple, the one we eat), the cultivar apple, was a popular traveler along the 4,000-mile ribbonlike trading paths known collectively as the Silk Route. When apple-eating traders traveled eastward and westward from Kazakhstan, they did the work of birds and bears by spreading apple seeds in both directions. Spengler, the archaeobotanist, writes that "by the time of the Roman Empire, the apple was revered from one end of the Silk Road to another."

We may know how the apple traveled to us, but botanists have heatedly debated the origins of the domesticated apple since the eighteenth century. The Swedish botanist Carl Linnaeus went as far as to argue that the domesticated apple be placed under Pyrus, the genus containing the pear. Earlier botanists strongly disagreed—as did British horticulturalist Philip Miller. The first to furnish a taxonomic description of the genus Malus, Miller argued that the apple should be placed within Malus, based on the different qualities of the pome—a technical term that refers to fruit with small seeds at the core, encased by fleshy mass and a thick membrane. Miller, however, didn't suggest which Malus species the apple should be assigned to.

In 1803, the German botanist Moritz Borkhausen provided the first positively defined description of the apple and named it *Malus domestica* (the name suggests a domesticated or cultivated apple).

Then Borkhausen threw a bomb: He conjectured that the apple was domesticated from the wild European crabapple, *Malus sylvestris*.

Scientists have passionately debated the conjecture ever since.

What complicated matters is that over a century later, in September 1929, geneticist Nikolai Vavilov traveled to Kazakhstan and had an epiphany when he encountered wild apple forests outside the city of Almaty. A founder of the Plant Institute, he surveyed the unusual apples—knobby yellow, green, and varying shades of red—assisted by a local teenager who would go on to become the apple scientist Aimak Dzhangaliev. Both Vavilov and Dzhangaliev noticed that there is considerable morphological or phenotype variation in *Malus sieversii*, the true wild apple. Some *Malus sieversii* trees even have thorns! On the whole, Vavilov was struck by the diversity of fruit size, shape, and color of the wild apples nestled in the Tian Shan mountains. Based on visual similarities between the fruits of *Malus sieversii* and the cultivar apple, Vavilov guessed that the genome of the cultivar apple was somehow linked to the wild apples he was seeing and theorized that this region near Almaty was the birthplace of the apple we eat today. Based on his observations, Vavilov came up with a compelling equation:

Centers of origin = Centers of diversity

"Centers of origin" refers to the geographical region where a fruit may have originated, and "centers of diversity" refers to where fruit variation has evolved over time. The theory was criticized as being overly simplistic, but genetic science would later prove that Vavilov was at least on the right track.

Reflecting on the Kazakh fruit forests, I find it thrilling to consider that the origin of the wild apple was not some anemic, barely-made-it event; instead, the diversity in wild apples points to the abundant creativity within nature. Interestingly, this diversity is also reflected in the structure of the wild apple trees, with trunks ranging from one-stemmed to five-stemmed to outright bushy, and crown structures varying from round to vase-formed to pyramidal.

Vavilov's theory would not become widely known until after the collapse of communism. A man of integrity and a brilliant scientist who corresponded with his Western peers, Vavilov soon grew out of favor with Stalin's regime. Cruelly persecuted for his independent mind, he would die of starvation in 1943 in a Leningrad prison. In time, his student Dzhangaliev would document the impressive diversity of *Malus sieversii* trees and fruit in their immunity to disease, biochemistry of the fruit, winter hardiness, quality and quantity of the fruit, ripening times, flowering period, and variation in tree canopy.

While many scientists came to believe that *Malus sieversii*, the wild apple that still grows along the flanks of Kazakhstan's Tian Shan mountains, is the primary ancestor of our cultivar apple, unanswered questions persisted. For starters, how did humans arrive at *Malus domestica*, the cultivated apple with roughly ten thousand varieties, simply from *Malus sieversii*?

The balance of morphological studies, which are based on physical features and traits such as flowering times, pointed to *Malus sieversii* as being the sole progenitor of the apples. But some scientists continued to wonder if *Malus sylvestris*, the wild European crabapple, might be a forgotten ancestral link between the wild central Asian apple and *Malus domestica*. Genetic evidence from the last twenty years has brought us closer to settling the issue. Newer and more sensitive genetic markers at last swung open the door for the debate to play out. Did other wild apple varieties contribute to the genome of *Malus domestica*? Could the flow of genes from crabapple trees actually have impacted the evolution of *Malus domestica*?

III

As I dove into the science of apple trees, I naturally craved to grow my own. Still, Michael and I were up against the impossibility of growing fruit at a time when the Southwest's megadrought, a significant indicator of a warming climate, had caused our clayey soil to become bone dry. It stung to recall the fruit trees we had acquired a decade back with enthusiasm from orchardist Gordon Tooley, who specializes in fruit trees that survive in our high desert environment. Historically, a variety of trees including apples, peaches, and apricots have thrived in our region. In our garden, Tooley's trees had received a royal welcome. Michael dug generous holes for the three hawthorns and a handful of fruit trees. That May, I was pregnant with Mia and my younger brother was visiting from Toronto. One afternoon, my brother and I watched through a glass door as Michael, swathed in bright sunlight, jabbed his shovel relentlessly against the unyielding ground. Turning to me, my brother remarked with satisfaction: "I am a city boy."

Despite the work Michael poured into transplanting and caring for the new trees, most of them remained stunted or even died. I still wonder if watering the trees more might have dodged failure. In an era of water scarcity, we use recycled kitchen water to supplement drip irrigation for edible plants. The Southwest is our home, and we save each bucket of water we can. Ironically, thinkers like William deBuys caution that saving water can also fuel the false notion that we have enough water to justify even more development; for when aridification really sets in, there will be no cushion as water conservation would already have hardened. In this scenario, agriculture, the highest water user in the Southwest, would be forced to cut back water use.

Whether or not there is water to support it, the drumbeat of development grows ever shriller. At a time when we can hear nature groaning, the country as a whole has seen a surge in development: Over roughly the last two decades, new development devoured 10 percent of America's land, mostly forests, equaling five times the size of Delaware. As nature bleeds away, a procession of birds, from beloved grassland species such as meadowlarks and northern bobwhites to common species such as warblers, finches, and blackbirds, decline precipitously. Which is not to say that we lack sensitivity toward birds—we adore birds, and so many of us love apples—but our sense of urgency remains blunted. The biodiversity crisis feels remote and unconnected to our lives. The word *crisis* itself is overused. How can we respond to anything seriously when so many places and people—our planet herself—are in crisis?

In a time of daunting upheavals, it is critical to *see* biodiversity loss, for truly seeing and experiencing can spur us to pivot and adapt to new climate realities. As a naturalist, I experience many upheavals in the natural world, and I make hard decisions about which ones to train my attention on. At the moment, the desiccated soil in my garden didn't look as though it could support the diverse variety of apple trees I was dreaming about.

Faced with a nearly dead crabapple, and long dead peach and nectarine trees, why did I dream about more crabapples, and even adding apple trees to the mix? An orchard is a space where I can taste and scent my childhood, for one thing. But growing the present is as important as the rasa of the past. On a quest to foster my connection to the wild, I know that crabapple trees, sour though their fruit might be, would be at home in our unruly garden. Biting into ripe apricots and apples straight from trees are among my juiciest memories from pandemic summers, a tangible connection to the biodiversity around us. Every state has its share of harvest festivals, but this way of being—with wild foods integrated into our everyday lives—is uncommon now. When my children offered freshly picked apricots to family friends visiting from a big city, the grown-ups stiffened and said, "Thank you," but didn't eat the juicy apricots.

At our local Audubon center, a seven-year-old boy mimicked Mia and climbed an apple tree. Mia has been scaling apple trees since she was five, and it is her fervent climbing that got us to notice the apple trees at the Audubon center. The boy, following her lead, also picked an apple. Then he raced over to his parents, excitedly showing them the apple and wanting to eat it.

His parents seized the fruit. "We have to think about it," they said.

It's almost as though we forget where the apple comes from.

Chapter Four

Seeking Celestial Apples

OCTOBER

A donkey friend at Las Golondrinas.

I

In October, chiles are roasted around town, and the chile-spiked air, scented with ripe apples and pumpkins, tugs at my soul. On the first Sunday of the month, I drive south for an annual harvest festival, which is held on a stop along an iconic trade route. Just as the Silk Route spread apples eastward and westward from Kazakhstan, arterial routes sprang up in America that would bring apple seeds and saplings into the deepest regions of the country, not only from England or France, but also from Spain. El Camino Real de Tierra Adentro, the Royal Road of the Interior, was the earliest Euro-American trade and supply route in the country. According to the Bureau of Land Management, the colonial wagon trail was a lifeline for Spanish travelers, who would stop at one last paraje—campground or resting place—before they journeyed on to Santa Fe, some fifteen miles upriver. That campground is now called El Rancho de Las Golondrinas, the Ranch of the Swallows, and I ritually stop here to look for the fork-tailed birds and their nests tucked into the porches of old structures. In the same way that the Silk Route transformed both the apple and how we eat, the Spanish who traveled on this sixteenth-century trail, which has coated my sneakers in dust, would alter the Western landscape not only with the introduction of fruit trees but also pack animals such as horses and donkeys.

In pre-contact times, Indigenous people used an array of wild foods—mushrooms, pecans, wild rice, purslane, and chokecherries. The Spanish added to this rich plate, but they seasoned it with grief as well. Spanish conquistador Francisco Vázquez de Coronado arrived here around 1540, and a seemingly endless line of Spanish explorers and friars followed on horses and carts and on foot. The Europeans brought diseases and exploitative ways to the region, along with members of the Rosaceae family—apples, peaches, cherries, apricots—and the melon, in the Cucurbitaceae family.

Walking in the direction of the pumpkin fields in Las Golondrinas, I cross a bone-dry creek—some years back, it had been so slushy and wet that a donkey refused to cross it and lay down sulkily in the mud despite the entreaties of a woman dressed in the full skirts of an eighteenth-century Señora. We'd been nearing the watery arroyo when we noticed two donkeys refusing to cross on the footbridge or by wading in muddy water. The stubbornness of Panchito, a brilliant white donkey, and High Beam, a gray miniature donkey, stirred glee within me.

High Beam was the smaller donkey, and his rider, in the chocolate-brown skirt, despaired—forehead lined and face darkened. Panchito's rider, in a dashing white tunic embroidered in red and black, also struggled to get his donkey to wade through the water. He gestured, sweated, and grunted, and only my children's presence prevented him from swearing. Unmoved, Panchito dug in his heels and then got deeply stuck in the clayey mud.

The children settled on the footbridge, legs dangling, to soak in the drama. The sun shone slyly through cloud cover, hinting at the thunderstorm expected later in the day. Barn swallows flew overhead, forked tails glinting. We were in the home of the swallows—and the donkeys.

Now it is a hotter and drier time, and I duck into the shade of an apple tree under which an old man and a young woman, dressed like Spanish colonial ranchers, are using a hand-operated metal press to crush apples for cider. They have extracted some juice in a long metal container, but they aren't sharing samples today because of COVID-19 restrictions.

In a white shirt embroidered with a flowery trim, the man dips a finger into the container and presses it to his lips. "It's really sweet," he tells me. He adds that the Spanish planted apple trees in this area when they first settled here, and subsequent trees were grafted from the original ones. "This is the oldest apple orchard in the United States—older than in Yakima County, Washington."

Wondering about the claim, I ask if I can taste an apple. He gladly offers me a green apple from a wooden vat where the fruit is waiting to be pressed. After thanking him, I walk on to an area where I crush grapes with my (sanitized) feet in order to make wine. My feet deeply purple, I wind back toward the pumpkin and chile fields. Feeling thirsty, I bite into the gift—the green apple. It is delicate, crisp, and subtly sweet. I am astonished. Other than a custardy apple I tasted long ago in a Himalayan orchard, I haven't tasted a more light-filled apple in my life. Eve may have been corrupted by the apple of knowledge, but this apple strangely buoys me.

II

Just before I climb "Steep Hill," as marked on the map, the path to my left forks into an abandoned dirt road edged with apple trees. I saunter over and discover canopies dotted with apples, and windblown fruit littered below. I take samples of the singular green apple I've tasted, along with a small russet-red apple whose skin has a dull sheen like the green's. Both apples are milk-white inside, and their taste transports me back to when an artist friend would brew Milk of Heaven, a spectral green tea with milky sweet notes. Biting into the shiny apples—the word comes from the German, apfer—I feel as exultant as when I first drank that otherworldly tea.

I scramble down a dirt slope to get to a couple of trees whose lower branches are heavy with fruit. There are so many apples on the ground that it's hard not to wobble over them while picking from the trees.

The sight of the ground layered with fermenting apples or apricots greets me in most places where I forage—and I always feel surprised that we let fruit go to waste in such quantities. The very fruit that was once so highly prized that it permeates our language with words and phrases such as "fruitful" and "the fruits of our labor."

"Only in America is fruit wasted like this," Jack Ortega, a fabled rose grower told me. "In Europe, where there's an emphasis on local food, this fruit would be harvested." Jack tends private historic gardens, and when ancient apricot trees yield an abundance of Spanish apricots—a juicier variety—he brings over a bag for my kids. He and his wife simmer the rest into a pastel orange jam that is scrumptious on toast.

Eyeing the wind-blown apples on the ground in Las Golondrinas, I console myself with the thought that they generally try to use or donate the fruit; importantly, the resident birds will also have fruit this winter. It comes to me in a flash—the singular bird I saw here one summer while walking along the pond. *Piranga rubra*, the summer tanager, redder than any apple, darting over the olive-green water, weaving through the leafy trees that rim the pond—a crimson apparition who easily overshadowed the bullfrogs croaking from the lily pads.

The sun blazes and, midway up the steep hill, I perch momentarily on a log and pull out one red and one green apple from my cloth bag. Both these apples have an ethereal taste that is more delicate than the dense, juicy Winesaps I picked last weekend at the Los Luceros Historic Site. A flicker flashes past.

The red bird, redder than dusk, had flown from one end of the mossy pond to the other. From the northern end of the moss-green water, three mighty bullfrogs glared. As the red bird whizzed past again, I could scarcely believe my eyes. The summer tanager's range is south of here, but it favors cottonwood groves, which might explain its preference for this celestial spot. Not only is the pond lined with wildly slanting cottonwoods, but the trees also luxuriate next to an acequia. It is not often now that New Mexico evokes the wet words of the Chinese poet Po Chü-I, but the pond came close.

> Masses of dark moss, earth newly rained on,
> Tiny drops of cold dew under an autumnish sky:
> don't stare at the bright moon, dwelling on the past—
> you'll only mar the features of your face, shorten your years.

On that summer-tanager afternoon, the children were quite small and, when they leaned toward the water to peer at the frogs, Michael worried they would stumble in. So, we decided to walk away from the pond. As we were leaving, the red bird dove past us, before my eyes, clear enough for three-year-old Pika to see. She turned to me, eyes shining.

"The bird gave us a goodbye wave," I said. This has been an enigmatic tradition in our lives: After a daylong trip to observe birds, when we're leaving a wildlife area, a remarkable bird often flies by smack in front of us. A goodbye wave. So to speak.

When I told a friend about the red bird, he said, "I was thinking about going to the pond. That gives me reason to go." I love that birds (and apples) give us reason to revisit a forest or a pond. All that evening and the next, the bird hung in my memory like a jewel. The unexpected flash of red in a sea of mostly green. The donkeys stuck in the mud would be remembered in my family for years to come. Being with them was like traveling back in time to a storied land of apples. I don't know where else the children might sit, feet dangling from a foot bridge, observing very stubborn donkeys being coaxed, nagged, and scolded out of squelching mud.

Back at home I give away samples of four apple varieties to friends or neighbors who stop by. A worker, Marcos, who helped fix a structural flaw in our pórtico, porch, is given a small bag of apples with the warning that some might have worms in them.

"They are tastier with the worms," Marcos says with a wry smile.

Marcos is originally from Mexico—the staging area where the Spanish conquistadores prepared for their long travels to what is now New Mexico. When I discover stray fruit trees in historic Santa Fe, I wonder if the heart of the city was built on the remnants of fruit orchards. Some four hundred years ago, the Spanish began moving northward into New Mexico on the Royal Road of the Interior, and along certain sections of the trail they planted the apple seeds they had brought with them. Clusters of saplings they tended later flourished as orchards. While many apple trees from that era eventually waned due to senescence or neglect, some mature trees persist, albeit in grafted form.

Research done by James Ivey, a National Park Service historian, suggests that the first apple trees in the Southwest were set in the ground as early as the 1630s by a Spanish rancher just north of the Salinas Valley; the area abuts the Manzano Mountains, which extend southeast of Albuquerque. "A large apple orchard was located on the site at the time the town (of Manzano) was founded in 1829," Ivey wrote in his report. "The settlers attributed the orchard to the missions of the 1600s, but it was probably in reality the property of Doña Salazar." A widow, Doña Catalina de Salazar was a prominent landowner, one of at least six farming and ranching families of Spanish origin who were well established in the area even at that early date.

"It should come as no surprise, then, that when colonists finally settled the region more than a century later, they named their modest community Manzano—apple tree in Spanish," another historian, William W. Dunmire, writes. "Ancient apple trees still grow there." The Manzano Mountains are a couple of hours south of Santa Fe, and their climate and altitude are said to mirror the finest apple-producing areas in Spain. According to the US Forest Service, in the early 1700s, Spanish explorers visited a small village on the eastern edge of the Manzano Mountains and discovered old apple trees there. Since apple trees are not native to this region, these must have been the trees the Spanish rancher (Doña Salazar's husband?) planted in the 1630s. A resident of the village of Chimayo, New Mexico, told me that one early Spanish variety was a small yellow apple, Manzana Mexicana, good only for baking, which they grow to this day.

It strikes me that the man in a rancher's costume who said that Las Golondrinas has the oldest apple orchard in the country didn't have his facts right. He would have been correct, though, if he meant to say that the orchards in this area are older than those in America's most famous apple region: Yakima County. Since the 1920s, Washington State has been the country's leading apple producer, growing over half of all our apples in Yakima County orchards, which span some seventy thousand acres. This large-scale apple operation began modestly when two of the Northwest's most notable missionaries, both energetic Jesuits, moved to Ahtanum in 1870 and 1872, as W. D. Lyman writes in *History of the Yakima Valley*. Father Caruana and Father Grassi were "among the Valley's first orchardists, setting out an apple orchard near the mission in 1872 which was irrigated by water from the Ahtanum Creek."

The Jesuits who seeded orchard culture in Yakima County were no doubt aware of other orchardists in their order, some of whom had brought apples to North America centuries earlier. The Jesuit order was founded in 1540 in Paris and formalized in Rome. Its adherents apparently lost little time in evangelizing overseas. By the late 1500s, French Jesuits are believed to have entered North America with apple seeds in their pockets. A later record confirms that by 1636 they were growing an orchard in Québec City, with apples, pears, and other fruit.

Like the Spanish, the French have a special place in North American apple history, having made some of the earliest attempts to grow apples here. The oldest capital city in the Americas north of Mexico,

Québec City was founded by French explorer and cartographer Samuel de Champlain. On a mission from France to search the Atlantic coastline for safe harbors and establish a site for the fur trade, Champlain became convinced of a "north-west passage" to the Orient and explored the interior of eastern Canada: the St. Lawrence river valley, the Ottawa River, and Georgian Bay (a vast bay on the eastern side of Lake Huron), as well as the lake that bears his name today—Lake Champlain.

Champlain's journals from this time and his *Narrative of a Voyage to the West Indies and Mexico in the Years 1599–1602* contain numerous observations on botany and gardening and reference strawberries, raspberries, plums, gooseberries, red currants, and the making of tortillas. While in Mexico, he noted that "there are apples, which are not very good, and pears, of tolerably good taste, which grow there naturally. I think that if anyone would take the trouble to plant these fruit in our climate, they would succeed very well."

Before founding Québec City, Champlain made several failed attempts to establish a settlement in present-day Nova Scotia: Isle Saint-Croix, during 1604–1605; then Port-Royale across the Bay of Fundy from Isle Saint-Croix, soon abandoned in 1607. At Port-Royale he befriended Louis Hébert, a colonist and apothecary from Paris. Hébert had created a garden at Port-Royale that included an experimental seed plot. When the settlement was abandoned, Hébert returned to France.

At Québec City, Champlain lost no time in planting a vegetable garden—"the first European experimental seed plot in North America"—which included beans, corn, squash, and vines, as well as cereal grains such as barley and wheat. The soil was untilled, and the plantings did well. A few years later, he met Louis Hébert again, this time in Paris. Praising the virtues of the land near Québec City, Champlain persuaded Hébert to settle back in New France. Hébert soon sold his home and garden and, together with his wife and their three children, left for Québec City in 1617.

History lurks closer to us than we imagine. My husband Michael's ancestors, on his maternal side, left France a few decades later to settle in the St. Lawrence Valley in Québec—a local plaque commemorates the family to this day. After a twist, I would later identify the main apple variety growing at Las Golondrinas and learn that the variety was established in America by way of a family from Québec!

"On the heights of Québec City," Champlain planted apple trees imported from Normandy and established an apple orchard as early as 1621. "On a cliff overlooking Québec," Hébert also planted apple and plum trees, which he had brought with him from Normandy. According to the testimony of a Brother Sagard, "the apple trees of Louis Hébert were covered with flowers in the spring, and in the fall brought good harvests." The apples were used for cider making, and Hébert's apple trees were still there in 1636, according to Brother La Jeune, though "spoiled by animals." It was enough to make La Jeune wonder whether apple trees would persist: "The apple trees that Lous Hébert had planted have been spoiled by animals, and it remains to be seen if the apples will become good in this country, on this I am rest assured."

Brother La Jeune's anxiety would turn out to be unfounded, and the cultivar apple would do resoundingly well in North America. But the apple would need many more helping hands along the way. Just as the apple's origin story has turned out to be more complicated than was once believed, the cultivar apple would take not one but many diverse paths into America. These lesser-known strands of the celestial fruit's history would come alive at surprising points in the course of my apple odyssey.

Chapter Five

The Fruitery

OCTOBER

Pika at the South Orchard, Monticello, Virginia. Photo by Priyanka Kumar.

I

We were at the kitchen table, the girls nibbling an afternoon snack, when Pika said, "A dollar a day keeps the doctor away."

I was so surprised that I couldn't speak for a full minute. I wondered at the way she had mangled the "apple a day" aphorism. The original, "Eat an apple on going to bed and you'll keep the doctor from earning his bread," came from Wales. It took a day before I could ask Pika about the origin of her saying.

She had picked up her dollar-centric version from classmates: "That's what they all say!"

Like generations before, Pika and her peers are dazzled by the glitter of consumerist culture. Today, they are hyperaware not only of shoes and games but also the ten thousand things that devices "feed" them, starting at eerily early ages. No doubt these kids will one day find ways to navigate materialism, as we did. Still, it can be a struggle for device-savvy youngsters, and parents similarly shackled, to cultivate an intimacy with nature.

At a climate workshop, I asked a veteran schoolteacher with a caring heart about her experiences with students. "It's hard to talk to young people today about the environment and its issues," Marty told me, chuckling uncomfortably. "They don't care."

Her response stunned me, and I told myself that it was one person's opinion. I trust that many kids do care. Nature is ideal stomping grounds for their keen sense of play and camaraderie. The apathy some kids might feel could surely have been stoked by the perception that older generations "didn't care" either—we passed along biodiversity loss and climate change to the kids growing up today. Since kids tend to mimic parents and caregivers, our own relationship with nature has the potential to inspire kids to care more. Such thoughts feel like lonely seeds, but Aeschylus once noted, "From a small seed a mighty trunk may grow."

We are our era. What will it take for us to want to grow apple trees for our grandchildren—as John Adams once did?

The apple is still supposedly our national fruit. Apples and bananas continue to be "the most consumed fresh fruit per capita" by Americans, though we now eat fewer apples than we once did. Since 2000, the consumption of fresh apples has fluctuated between fifteen and nineteen pounds per consumer each year. The most recent numbers show that we eat some fifty apples a year, less than one a week. After a barrage of insipid and oversweet apples, it is perhaps understandable that our apple love has dried up. It's as though we sense that the apple is blemished at its core. Our current disconnect from orchards exemplifies how our relationship with the land has frayed to the point that new generations are more conversant with Apple (Mac) products than apple varieties.

When I speak to friends and neighbors in their sixties, seventies, or eighties, many have delightful memories of playing in neighborhood orchards. Amy grew up in Minnesota and recalls after-school wanderings with friends that ended with her standing on a pony to pick apples. Dwight grew up in northern New Mexico nearly eight decades back and recalls getting chased by orchardists and rebuked for "stealing" apples. His parents were schoolteachers and owned a farm and ranch in rural Colfax County, but they abruptly sold it while he was away in college.

"I still dream about our land," Dwight says, when I see him on my neighborhood walks. "I was born there, on a bed, without a doctor. . . . At Christmastime, we kids would go and find the right tree, cut it down, and bring it back to the house." His eyes gleam as he recalls tree-filled days. "After all these years, I still think about how I might have stopped them from selling the farm. Some days, I wake up thinking about that."

In the twentieth century, the loss of small farms became inexorable, and the cookie-cutter urbanization that followed continues to spread like a rash over the land. We no longer have tree-filled lives, and kids have fewer opportunities to cultivate a lasting affection for the land as Dwight and Amy once did.

I know Amy by way of a friend, Richard, who grew up in Washington state—the land of apples. Over the years, we've had a running conversation about rasa. "I miss the apples of my youth," Richard said one day. "Apples today don't have the same juice."

"D'you mean to say juice, as in apple juice, or that apples are lacking rasa?"

"Exactly! Rasa! And a lot of people have lost it too. . . ."

I smiled ruefully. If there's anyone who has knit together a community for himself, it is Richard. It is what keeps him thriving at the age of eighty. But he pours time into fostering community in the same way that I make it a practice to foster a connection to nature.

II

It is perhaps not surprising that since its infancy, America has loved apples. There's something sturdy about the apple, a fruit that thrives in temperate regions and is reminiscent of a tart-and-sweet grandma. One might fall on occasion for the sublime peach or the succulent fig, but one unfailingly returns to the bread-and-butter apple. America's love affair with apples can be parsed into roughly three phases. The first phase begins with early apple plantings by settlers and Spanish and French colonialists in the 1600s and 1700s. The next phase brought the impassioned efforts of visionaries such as John Chapman and Jean-Baptiste Lamy in the first three-quarters of the 1800s. A third phase of American apple infatuation began when white settlers dreamed big about apple orchards, capping what the National Park Service calls the "golden age of fruit growing in the 19th century," as Susan Dolan writes in *Fruitful Legacy*. According to the US Department of the Interior, by the end of the 1870s, commercial and fruit orchards could be found throughout the forty-eight contiguous states. These commercial orchards, some of them passion projects, would later give way to industrialized orchards—the fourth phase we now find ourselves in.

While the Spanish brought apple seeds to the Western US by way of Mexico, settlers from countries such as England and Ireland also traveled to their new home with apple seeds. By 1623, they were growing seedling orchards in New England, primarily for cider. Intriguingly, a Jamestown resident named Ralph Hamor mentioned an orchard there as early as 1615. Sir Thomas Gates had built his residence next to the governor's mansion, and Hamor writes that "many forward apple &

Pear trees come up, of the kernels set there the yeere before" in Gates's garden. American apple culture was taking root, and "as American as apple pie" would go on to become a time-honored phrase (although the original recipe came from Britain). During the domestication of the apple from its *Malus sieversii* progenitor and through its hybridization, the apple has been under continuous selection for fruit size, flavor, resistance to disease, keeping quality, firmness, and sweetness. The migration of the apple to America added a new chapter in its domestication, with many American cultivars believed to have arisen from chance seedlings (a rare event when a superior apple variety sprouts from seed). Settler orchards became excellent labs for new varieties and introduced fresh genetic diversity to a genetic pool that is otherwise generally frozen when the propagation of cultivars happens solely through grafting.

For the first 150 years of our apple history, American farmers regarded grafting with dark suspicion. To take a limb from one apple tree and glue it to an alien trunk appeared to be an unnatural, even devilish act. But cloning and grafting are the only ways to grow specific apple varieties. Apple seeds are heterozygous and reproduce sexually. This means that a large portion of the apple genome preserves genetic information from *both* progenitors. So, simply planting an apple seed will not give you the apple variety you covet. That being said, grafting apple trees is an enterprise riddled with failure—as I would find out when I started grafting to deepen my land fluency.

Fruit-bearing trees with desirable qualities would have been vegetatively propagated—reproduced asexually—before the invention of grafting. The first fruit trees domesticated in the Old World, during the middle to late Neolithic period, appear to be the grapevine, olive, date palm, and fig. Both grape and fig can be vegetatively propagated by removing a twig by a cut and then planting it into the soil. Date palms can be propagated by transplanting shoots; for olives, cuttings suffice, though it is perhaps not the best way. Apples and other fruits in the rose family such as cherries, pears, and plums, on the other hand, cannot be vegetatively propagated by these simple techniques, and so it wasn't until the invention of grafting that the domestication of the apple really took off.

Johnny Appleseed famously did not believe in grafting. He got away with relying on apple saplings because in early America, when he practiced his craft, apples were primarily cultivated for cider; tannic and acidic varieties were also in demand for hooch. Cider mills were everywhere, and root cellars brimmed with apples, which were also deployed for baking, drying, saucing, and blending into brandies. The first important varieties used for cider and dessert in America are from roughly 1650: the Roxbury Russet, from the Boston area, and the Rhode Island Greening, both with complex flavors that would be considered highly acidic today. Our early economy was cash poor, and barrels of hard cider and applejack were commonly bartered, even serving as payment for our first road-building crews. The tradition of paying part of a farm laborer's wage in cider continued in Britain until 1878, when it was made illegal. American voters were a demanding lot—they expected to drink hard cider before going to the polls. In 1755, George Washington refused to "swill the planters with bumbo"—soften his electorate with hard cider—and lost the election to the Virginia House of Burgesses. Three years later, his campaign distributed some 150 gallons of hard cider, and he won easily.

In the eighteenth century, apple trees were seemingly everywhere, and American orchards had an enviable diversity. Thomas Jefferson grew a whopping 170 varieties of fruit trees in his orchard at the Monticello plantation. His 1778 plan drawing with meticulous notes of his "fruitery," as he called it, is believed to be the most detailed surviving record of an eighteenth-century American orchard. Jefferson, Washington, and Adams craved to move away from highly curated British gardens and invite at least a taste of unruly nature into their own gardens and orchards; and they all but dreamed about fruit trees. Jefferson focused on cultivars that were ideal for cider production. He likened the Hewes Crab, with its green dotted skin and yellow flesh, to a sponge, dripping with juice. The Taliaferro was "the best cyder apple existing ... nearer to the silky Champagne than any other." For dessert fruits, he favored the iconic Esopus Spitzenburg, a 1790 variety from Esopus, New York, whose "aromatic flavor" improves "radically" with storage and is at its finest at Christmas, and the hardy Newtown

Pippin, which is believed to be among the oldest varieties in the country, dating back to the 1750s.

I wanted to visit Jefferson's eight-acre fruitery in Monticello to unpack how the land was his laboratory and fruit growing an essential component of his "simple farmer" image. Jefferson, of course, was a terribly complicated man. The fruit he grew depended on the labor of at least some of the roughly 130 enslaved people who, according to the Jefferson Foundation, lived and worked at Monticello at any given time—in addition to a hundred "free workmen" employed during Jefferson's ownership of the plantation. In theory, Jefferson opposed slavery—early on, he despised the system but couldn't see a way out from an economic standpoint. He preferred to give his slaves "incentives" rather than "disincentives," but he wasn't above having one runaway slave flogged. After he became a widower, he went on to father kids with a slave, Sally Hemings, who was also his late wife's half-sister; so, enslaved people were literally part of his extended family. Another enslaved woman, Ursula, who nursed Jefferson's oldest daughter, Martha, was the rare person whose skills he trusted to distill cider each year from the apples he grew. Fruit growing and cider making are undoubtedly labor intensive, and Jefferson's fruitery, household, and wheat fields all functioned on slavery's back.

The past isn't always palatable and may contain seeds of what is abhorrent. In researching Jefferson's orchard, I struggled with his slave-owning past and his relationship with Hemings. Perhaps the only way to approach the past is to do so warily, the way a teenager approaches her parents. Washington state may have a $4 billion apple industry today, but it was born on "the traditional gathering grounds and reservation lands of the Columbia River tribal peoples: the Chinook (at Fort Vancouver and the Cowlitz Farm); the Nez Perce at Alpowa (Chief Timothy's orchard), and at Fort Simcoe (Yakama Indian Reservation)," as Mary Rose writes for the Confluence Project. Across the country, Indigenous peoples and African Americans labored to sustain young orchards.

The abundance of cheap labor allowed apple orchards—and their diverse varieties—to flourish nationwide until the nineteenth century, when apple diversity began to decline. But did the loss of cheap labor contribute to this decline? If that were the case, it still wouldn't entirely explain why thousands of apple varieties have gone extinct since the twentieth century, including Jefferson's beloved Taliaferro. Biological factors, including the practices involved in modern monoculture orchards, may be partly to blame. In Jefferson's era, orchards routinely grew a diversity of fruits, which was also a form of insurance against diseases that certain varieties might be susceptible to. There were no synthetic fertilizers, and orchards were instead fertilized with manure. Jefferson favored wild undergrowth, which encourages beneficial fungi and soil richness, whereas industrial fruit growing deploys herbicides to kill this very undergrowth. Why does undergrowth matter? The botanist Tom Antonio told me a story about Central American cacao growers who found out that their cacao flowers weren't being pollinated. The growers eventually brought in ecologists, including Allen M. Young, to study the problem. Young found that because there was no undergrowth left, there weren't any places where the midges, who pollinate cacao flowers, could complete their life cycle. One of Young's most interesting discoveries was that adding piles of cut banana stems under the chocolate trees provided a moist refuge for the midges even during the driest weeks and helped maintain the pollinator population. He advised the growers not to obsess about on-the-ground cleanliness and let leaf litter alone. A season later, the midges made a comeback and the cacao flowers were once again fertilized.

Jefferson similarly favored nature-centered solutions. When he read that seaweed discourages pests, he dispatched the information to the editor of a gardening journal so that fellow Virginians might also experiment with seaweed. John Adams, too, was into seaweed. In a diary entry on September 5, 1796, Adams noted that it was "The Anniversary of The Congress in 1774" before recording how he spent the day: "Sullivan brought a good Load of green Seaweed, with six Cattle, which We spread and limed upon the heap of Compost in the Meadow. Carted Earth from the Wall to the same heap." Adams staunchly opposed slavery and never owned slaves, and it's refreshing to picture him working alongside a hired man to improve a compost pile. Engagement and experimentation hone our intuition about the land. Jefferson grew figs in variable areas created by stone and shade, which makes me wonder if he had an intuitive knowledge of microclimates.

Taking his apple love to heart, Jefferson was dismissive of European varieties. He wrote home from Paris, "They have no apples here to compare with our Newtown Pippin."

Benjamin Franklin was more tactful. He used the complex taste of the Newtown Pippin—a variety that I am now growing—as a diplomatic tool to win over the British. This green apple with a sunny highlight is an excellent keeper, and its tartly fresh taste grows more complex with age. The Newtown Pippin became such a smash in Britain that the country began to import it in large quantities. Later, Queen Victoria herself became a devotee. Leave it to Franklin to score an economic, political, *and* cultural win.

In the eighteenth century, many of Virginia's gentlemen farmers fervently grew apples. After leading the Revolutionary War against the British for seven long years, Washington returned home to Mount Vernon, weary of public life. He began to design the tree plantings on his estate with military fervor, not even waiting for the ground to thaw before he ordered his men to dig holes for new transplants. The next spring, many of Washington's young trees bloomed enticingly, only to deteriorate and die. Unfazed, he and his men planted a new battalion of saplings that fall. Once again, there can be no doubt that enslaved labor kept Washington's garden and kitchen running. We know that toward the end of Washington's life, his enslaved workers included three gardeners—George, Harry, and Joseph—who worked under a hired English gardener, William Spence. In his will, Washington freed the enslaved persons whom he legally owned, including George, one of his gardeners. George's wife, however, had come into the family through the estate of Martha Washington's first husband and remained enslaved.

Martha Washington liked the garden to produce a bounty of vegetables for her family and their guests. Beyond the lower kitchen, the Washingtons planted a diverse range of fruit trees, with orange trees planted, interestingly, in the ornamental garden. The family frequently hosted dinners, and it wasn't unusual for Martha to stop by the kitchen four times a day to oversee arrangements. She had inherited thirty-eight-year-old Doll from her first husband's estate and brought her to Mount Vernon. Doll cooked for the Washingtons and prepared their weekly social dinners, sometimes putting in fourteen-hour days, beginning at four in the morning. The day's main meal was served at 3:00 p.m. The first and second courses were meats and vegetables and then the tablecloth was removed for the third course of nuts, raisins, and apples, served with port and Madeira wine. A wine glass in hand, George Washington would hold court over the male guests while the women drifted to the living room.

As Doll grew older, she trained her daughter, Lucy, to take over as the head cook. In her sixties, Doll wove and mended like other female slaves her age—but the Washingtons still expected her to distill "a good deal of" rosewater and mint water, in addition to preserving cherries and gooseberries. An astonishing new discovery gives a glimpse of the handiwork of Doll, or a contemporary of hers. Thirty-three bottles of preserved cherries and two more bottles, possibly containing gooseberries or currants, all standing upright, were uncovered by archeologists studying an old cellar site in Mount Vernon, ahead of a planned restoration. When some of the dark green glass bottles were opened, the cherry fragments inside were found to be extremely clean and still gave off the fragrance of blossoms—some 250 years later! Whether Doll or another enslaved woman preserved the cherries, her fruit-preservation skills were exemplary.

Knowing fruit, and what drink it yielded, was essential to living and savoring life. John Adams wouldn't dream of starting his day without a "jill" of hard cider. From his time as a Harvard undergraduate to the end of his life, Adams drank "refreshing and salubrious" hard cider before breakfast. He and his wife, Abigail, grew their own apples and pressed their cider on a farm in Quincy, Massachusetts. Hard cider, with British roots dating back to at least 55 BCE, was a beloved American drink, not least because it was safer than unclean or brackish drinking water. When fermented, hard cider turns into an elixir, apple cider vinegar, used as a food preservative and pickling agent to make unforgiving winters more palatable.

When traveling in the country, John Adams relished tavern food, including fresh fruit. He went on to serve as the first American ambassador to Britain, but he was happiest rooting about in his garden near Boston, pulling out weeds or planting an apple tree. It says something about the kinship between him and Abigail that she once confessed to their mutual friend, Jefferson, that she preferred to tend to her garden than participate in the doings of the British court. Before their departure from Britain, she wrote to Jefferson that she couldn't wait to return home: "Improving my Garden has more charms for my fancy, than residing at the court of Saint James where I seldom meet with characters so inoffensive as my Hens and chickings, or minds so well improved as my Garden."

My heart lifts at the thought that Abigail saw her garden as having a mind.

The founding fathers, and mothers, were passionate about apples, though they may not have known that the fruit is loaded with vitamins (notably vitamin C), fiber (pectin), minerals (potassium), antioxidants, and flavonoids—particularly quercetin—a group of molecular compounds researchers say are beneficial in the prevention of cancer and other conditions. Adams may not have known that apples reduce bad cholesterol and the risk of strokes, but he was religious about his cider habit and saw the apple as a health-giving fruit. In his diary, he wrote: "In conformity to the fashion I drank this Morning and Yesterday Morning, about a Jill of Cyder. It seems to do me good, by diluting and dissolving the Phlegm or the Bile in the Stomach." Preferring the cider from his own orchard to any other, in another diary entry he expressed the hope that "in the twentieth century...my Grand-children *may* live to see, an Apple-tree from a seed planted by my hand."

I similarly wish that my girls, and their generation, will one day harvest an abundance of rich apple varieties. In our garden, which certainly has a mind of its own, the girls skip over rocks and skirt abundant cacti while glancing at Michael and me tending to our young fruit trees, against many odds. If we can keep nourishing this soil, with some luck we will soon be growing apple trees with our own hands.

Chapter Six

Ten Thousand Varieties

OCTOBER

A feral apple. Photo by Priyanka Kumar.

I

It was perhaps inevitable that my apple wanderings would lead me to an ingenious grower of apple trees—a largely self-educated man. Although a small number of commercial apple varieties dominate supermarket shelves today, a few stubborn orchardists across the country keep historic varieties in circulation, thus oxygenating America's apple traditions. I had met Gordon Tooley ten years back while acquiring peach, nectarine, and apricot trees for our garden, but this time our encounter was different. On a bright October morning, my family and I drove an hour north to Tooley's Trees to look for interesting varieties that might thrive in our garden. On the way north to Truchas, where the farm is perched, maternal cottonwoods blazed yellow-gold, like nature's torches. The undulating hills all around were squat and clay-colored with only desert scrub for vegetation. An epic sparseness lit the land. Dark gray clouds sulked in the sky until the sun winked through and gilded the landscape. Tipped with shimmering snow, the Truchas peaks shone against the adjacent mountains, which bloomed yellow as though smeared with wildflowers.

Michael abruptly turned into a dirt road without any signage and drove through a small gate with a nondescript sign that read Tooley's Trees. Glancing at the odometer, I noted that our Honda Accord hit 100,000 miles as we drove into a flat expanse that was both orchard and nursery. Our sole vehicle, the Honda has been mostly faithful, but as it bumpily traversed the primitive road into the orchard, I hoped that the axle would stay intact for the ride back home.

Garrett, a young man with a toothy grin, walked us to a grassy area where young apple trees were huddled, rooted in white fiber bags. "Gordon's root bags are robust like him!" a devotee of Tooley's Trees would later tell me. It was nearing the end of the season, and only one Hewes crabapple was left. It is unknown exactly when the Hewes originated, but Hewes trees were discovered in Virginia in 1817 and were then believed to be a hundred years old. The tree produces a fantasy fruit with dark green skin, which Tooley's catalog said would get covered with purple-red and many large white dots. It sounded very Alice-in-Wonderland, and I was in.

To Garrett's delight, we also picked up an Albemarle (Newtown) Pippin, though the catalog warned that this apple doesn't owe its success to good looks. The Newtown Pippin is instead first-rate for desserts and cider, and an excellent keeper—which was an important consideration when the founding fathers exalted it, and young America exported it with resounding success to Britain.

An iconic apple deserves to bring along an offspring—the Wickson crabapple ("more crab than apple"), introduced in 1944, is a cross between the Newtown Pippin and the fabled Esopus Spitzenburg, considered to be one of the finest varieties ever known. Here is some apple math to fix the genealogy:

Newtown Pippin (Jefferson's favorite) x Esopus Spitzenburg = Wickson crab (1944)

To our threesome, we added a final companion: a two-hundred-year-old Claygate Pearmain apple, "found growing in a hedge in the hamlet of Claygate in Surrey before 1820." Four trees were a handful, but I found myself eying the description of the Charette apple, which may have been brought to Fort Kent, Maine, in the eighteenth century by French missionaries—brothers of Santa Fe's legendary fruit grower, Archbishop Jean-Baptiste Lamy. But the Charette was "out of stock."

After many adjustments, which felt like solving a puzzle, we fitted the four young trees into our car. Perhaps if these young trees were to become bountiful, we'd need our own fruit cellar. It was hopeless to dream in this way, but I suppose I was fruit-starved enough to indulge the fantasy.

Garrett walked us over to Gordon Tooley, who was kneeling on the ground in the thick of the ten-acre farm, his intense blue-eyed face fringed with white hair and a snowy beard. He was fielding questions from a man whose mother lives in a six-hundred-year-old house that was once part of the Pojoaque Pueblo. The man had brought apples from their historic orchard for identification, and Gordon had spread out a handful on the ground. Cutting an apple with his pocketknife, Gordon offered me a slice. The verdict wasn't as intriguing as the

Pojoaque man had hoped: Red Delicious and Winesap—and both varieties showed signs of the coddling moth.

"The five knobs on the bottom are a giveaway," Gordon said about the Red Delicious. "The original name of these used to be Hawkeye." The classic Delicious shape is a round apple that becomes slightly longer and tapered. While it's the country's most widely grown red apple, making up a whopping 41 percent of apple production, the flesh of the Red Delicious can too often be cardboardy and the flavor tepid. Who can blame generations of schoolchildren for giving up on apples after swallowing the Red Delicious day after day for lunch? Before the Red Delicious grew ubiquitous, the Winesap, an old apple that originated in New Jersey before 1800 and became a popular Southern apple, was a "heavyweight," famous for its twang and cider and prized as a glorious keeper; when researchers learned how to pause an apple's ripening in an atmosphere of nitrogen gas, Washington's apple industry abandoned the complex Winesap for good-looking varieties such as the Red Delicious.

My decade-old acquaintance with Gordon was soon renewed and, after the apple tasting, he gave me a tour of his interesting grafts. As we walked through his orchard, I wondered aloud if he might care to take a field trip to the Real (pronounced *Ri-al*), one of the oldest orchards in the area. Over a decade back, a Bureau of Land Management employee, intrigued by the Real's history, encouraged the BLM to snap up the property when it became available. Now that employee has moved on, however, and the orchard's future has grown uncertain. Some years back, when the Real fruit trees began to decline, the BLM tapped Gordon to map and monitor the orchard. While caring for the Real trees, Gordon took scion wood (the previous year's fresh growth) from trees that were failing in order to propagate them at his farm—so that the varieties wouldn't be lost to the world. Walking me to the site of the grafted trees, Gordon said: "The trees on the west side [of the Real property] are gone.... The ditch blew out on the Santa Fe River and they didn't get watered anymore."

The historic fruit trees were already neglected when an extreme "one-in-a-thousand-year" storm disrupted an ancient acequia, and the water-starved Real trees—apple and pear—are now dying or dead.

Once-thriving orchards across the country, including those in Yakima County, are similarly facing climate events. The unfortunate fate of many of the Real trees is reflected in the names that Gordon and his interns gave the varieties they propagated: Nearly Dead apple; Nearly Dead pear; Meat Hook pear (the tree had hooks for hanging a hog upside down); Mama pear, the largest pear tree on the property, which is now dead. It pricked me that half of the old Real fruit trees had perished from the storm and neglect. Gordon and I agreed to check up on the surviving Real trees in a couple of weeks. To take a breather from the gloomy history of those trees, we walked on to a beautifully wide and robust Kazakh apple tree, which Gordon grew from a seedling.

"There's this missionary guy who brings seeds back for me," he said, chuckling. "He sews the seeds into the lapels of his collar."

It is a stroke of fortune that wild apple trees still grow today in the Tian Shan, the celestial mountains, in southeastern Kazakhstan. Development has caused even these remote, ancestral apple forests to shrink in size, however. As Gordon showed off his Kazakh tree, a dream flared inside me—to hike through *Malus sieversii* forests and see the ancestor of today's cultivars. There, one might truly feel like Alice in Wonderland.

There are others who bring Gordon treasures: apricot seeds from Ladakh, India, or twigs and seeds from Iran, where many varieties of fruit trees are losing their habitat. "This is what we could do," I thought, envisioning a network of fruit lovers working across borders to maintain biodiversity in fruit. The apple historian Dan Bussey has estimated that there once were more than sixteen thousand named apple varieties in the United States alone. Historically, apples had more diversity than any other crop domesticated in America or introduced to the continent. Astonishingly, we have now lost more than half of those varieties. Apple conservationists believe that only three thousand varieties remain accessible to us: An estimated four out of five apple varieties unique to North America have been lost, in the sense that they are no longer in circulation.

"No genus can afford to lose that much variety," Gordon said.

I wondered why thousands of scrumptious apple varieties have gone extinct or are entirely forgotten. In our everyday lives, we rarely encounter more than a handful of varieties. According to the Renewing America's Food Traditions (RAFT) Alliance, eleven varieties produce 90 percent of all apples sold in chain grocery stores. It almost seems as though someone had decided to jettison the interesting old varieties. It turns out that isn't far from the truth. Researchers now know that conventional apple breeding has "long concentrated on a very small number of cultivars."

Still, why aren't historic orchards with rare varieties protected in the way that old-growth groves sometimes are? There have been efforts to identify mature orchards that stand within the boundaries of national parks—amazingly, 34 percent of national parks have historic orchards—but those initiatives may have sputtered. The RAFT Alliance estimates that 81 percent of all apple varieties available to gardeners, orchard keepers, chefs, and cider-makers are endangered. Perhaps rare heirloom varieties of fruit trees on public and private lands should be offered similar protection as endangered trees or birds.

The apple isn't the only fruit whose rich varieties are languishing. The banana boasts more than a thousand varieties, "each with its own genetic variation, color, shape, and size," as Alexandre Antonelli writes in *The Hidden Universe*. But a single variety, the Cavendish, named after an English duke, accounts for half of the world's banana production and 99 percent of all exports. Perhaps it isn't surprising that a virulent fungal strain is spreading like fire through the continents where the Cavendish is grown and blooming into a "pantropical disease." Scientists expect the blight to impact most monocultures of the Cavendish banana—so, those farmers who grow a higher diversity of varieties will be much better off. While gene banks stand ready to tackle catastrophic scenarios, Antonelli cautions that "we place ourselves in an incredibly vulnerable position by relying on so few species, since a single pest or pathogen could rapidly wipe out vast plantations." Such a prospect is not science fiction; Antonelli cites Ireland's Great Famine in 1845–49, when "a fungus-like organism destroyed the potato crop, the dominant food source for much of the population at the time."

As we go on losing feral and rare varieties of fruit trees, we are swimming in a perilous sea of sameness—in the "holy trinity of Fuji, Gala, and Braeburn," as Gordon acidly says.

The limited varieties grown in the system of gridded monoculture may produce big-and-bright-looking fruit, but as climate challenges mount, the trees are becoming more vulnerable to environmental stressors and devastating diseases. One of these diseases, called rapid apple decline, is spreading across the country. RAD causes apple trees to collapse en masse and has confounded scientists. Its costs are now becoming clearer. "The impact of RAD is estimated to cost Pennsylvania apple growers approximately $13,000 per acre and New York apple growers $3,380 per acre, which is a heavy financial burden," says Ruhui Li, a plant pathologist at the National Germplasm Resources Laboratory in Beltsville, Maryland. New insights from Cornell University scientists confirm that steering away from overcrowded monoculture orchards may be the only reliable way to fight diseases such as RAD.

Diverse orchards also better survive climate challenges such as sharply reduced "chill hours" and extreme temperature fluctuations. Our dependence on commercial varieties—and inexpensive, passable apples—comes at a steeper price than we realize. As we grapple with climate catastrophes, the need for genetic diversity accrues urgency.

"None of this belongs to any of us," Gordon added, referring to the Real varieties he has propagated. He plans to give some away to schoolchildren next summer. "I would rather take twigs from Real than wait for an act of Congress to allow me to do so." The twinkle in Gordon's eye acknowledged that, legally, he wasn't supposed to take twigs from Real.

"In this case, you did the ethical thing," I said. Saving old varieties from a dusty death and sparking new life into them is a noble endeavor. Gordon had to tap into his inner Robinhood to save the Real varieties. Scientists warn, though, that we must remain cautious about viruses that old varieties might be carrying.

We walked up to a lively Baldwin tree and tasted its large red apples. A man named Loammi Baldwin popularized this variety in the late 1700s after he noticed woodpeckers swarming about a tree that bore bright red winter apples. By some coincidence, Baldwin was the best friend of a cousin of Johnny Appleseed. Biting into the juicy fruit, I was struck by the exceptional, sparkling flavor. I called my children over.

They had spent this time racing through the apple trees, and six-year-old Pika swung over to catch an apple before sprinting away.

Observing her, Gordon nodded. "It's good you let them run around. . . . Most children who come here don't even get out of the car."

I eyed him, feeling stunned. I gestured to the open expanse of the orchard and mountains, and the two kids weaving through the trees. "I'd like them to grow up like this."

Gordon told me that he had grown up "like this" in Cimarron, New Mexico, and we were soon comparing notes. He and I grew up in different countries, it is true, but our childhoods had a flavor in common: Once we returned home from school, we were allowed to run off and play in nature—crucially, nature still existed and even flourished in our neighborhoods—and we weren't expected back before dinner. It struck me that it's not just apple varieties that have been lost. A way of living and being in nature has also become endangered.

A little later, as we were about to drive away, I saw Gordon hurrying over to our car. I opened my door.

"Do you use tomatillos?" he asked.

"Michael makes an excellent salsa," I said.

Gordon was carrying a white bucket crammed with emerald fruit, and he emptied dozens of tomatillos into an empty tortilla bag until it was full. Handing me the bulging bag, he waved us goodbye.

II

As we drove back to Santa Fe, I was aware of an effervescence in my heart. There was something gnome-like about Gordon; he had understood my love of apples and winked back at me. While my apple love got me brooding over the loss of so many rich apple varieties, and the need to preserve genetic diversity, I also found myself thinking about the children Gordon had spoken of, those who don't get out of their cars in an apple orchard. My father had taken me to the Himalayan orchard when I was roughly five. What would I have lost if I'd stayed in my father's jeep instead of inhaling the tart-and-honeyed aromas of apple trees? Some of the phantom images that I can recall are of me walking behind him through a grove as he questioned the orchardist about pruning techniques, weather patterns, and the varieties being grown. Later, he had understood my delight over a custardy apple, and somehow these intangibles helped foster a love for apple trees that persists to this day. I cannot quantify what would have been lost had I stayed in the jeep, but what was gained contributes to the essence of who I am.

I wonder how many children get such foundational experiences today. Pika tells me that several kids in her first-grade class have television sets in their bedrooms and when they wake up, "They right away press the remote to turn the TV on." Some children keep their TV sets on for hours, and a subset of these six-year-olds have their own smartphones. These kids linger within the four walls of their rooms and cars, and childhood wanderings tend to take place in stores, competitive sports, or inside a device. My own kids will no doubt consume some share of technology, and Mia has already asked about a phone (most of her elementary school friends have one). For now, we have decided to wait. I would rather give her experiences that are greenly resplendent and take her along on wild foraging trips. When I witness the harried landscape that kids navigate today, I realize how fortunate I was to absorb the thrill of meandering through forest-like orchards, laden with awe-inspiring fruits of staggering colors and taste. I wish I could share the joy of jumping out of my father's jeep and meandering all day among apple trees. The journey that began that day in the Himalayan foothills has fed an underground spring and nourishes my desire to deepen our connection to fruit culture. If I weren't sipping from that spring, I would watch this land being lashed by wild storms and temperature fluctuations and soon grow disheartened.

Now, as we drove back home from Truchas with two children and four trees in tow, I could feel the trees shifting in the trunk and back seat and wondered whether our elementary schoolers would grow up in conversation with fruit trees. Michael also had an über-outdoorsy childhood, canoeing and portaging in the summers in Algonquin Park and the Temagami area farther north in Canada, but I was astonished to discover that he had never picked fruit from a tree as a child. Over a dozen years back, before we had kids, we visited the farm of a woman who sold lovely dried flowers at the farmers' market. At the end of the farm tour, we lingered, and she came over and told us that her white peach trees were making a lot of fruit. "You are welcome to pick what you want." The late afternoon light gilded the abundance of flowers in the farm, with black-eyed Susans, lavenders, and amaranths all but glowing as we stood under a prolific white peach tree. We picked deeply fragrant fruit and ate it right there. Cocooned by the dusky light, live bouquets of flowers, and the rasa of peaches, I almost fainted when Michael told me that this was the first time he'd eaten fruit straight from a tree.

Chapter Seven

The Flow of Energy

OCTOBER

Planting a tree for newborn Mia.

I

Back at home, we planted all four apple trees. Our previous attempts at growing fruit trees had not been successful, and this time around I was anxious but excited to finally bring my love of apples home. The Wickson crabapple went near a wooden swing on the eastside, where it might one day shade a swinging dreamer. It felt essential to plant a crabapple, if only because it has emerged as an unexpectedly important character in the apple's story.

While consensus grew among scientists that *Malus domestica*, the cultivar apple, ultimately originated from *Malus sieversii*, the wild Kazakh apple, the story of the apple's domestication would turn out to be more circuitous than what was once imagined. The domestication of fruit and seed crops typically follows two phases—first, an unconscious selection of plant types occurs, followed by a more consciously directed artificial selection for specific characteristics. In the case of the cultivar apple, three crabapple species emerged as the strongest contenders for gene-flow contributions: *Malus sylvestris* in Europe, *Malus sieversii* in Xinjiang and Tian Shan, and *Malus orientalis* in the Caucasus.

Beginning in the early 2000s, comparisons of DNA types between species revealed a more nuanced and complex path to apple domestication than what was previously believed. The biggest surprise from early studies of the chloroplast DNA was that a considerable measure of the apple's chloroplast is inherited from the European crabapple. So, the German botanist Borkhausen had been prescient when he dropped this particular bomb in 1803.

Recent studies continue to make tantalizing advances. A group led by the French scientist Amandine Cornille looked into whether other wild apple species along the Silk Route could have contributed to the genetic makeup of *Malus domestica*. Their study used microsatellite markers and an unprecedented sampling of 839 accessions from five apple species from China to Spain.

The findings sharply upended the long-standing belief that *Malus sieversii*, the wild central Asian apple, was the sole ancestor of *Malus domestica*. Cornille's team showed that not just one or two, but multiple apple species contributed to the genome of our domesticated apple—and the wild European crabapple, which diverged from the central Asian apple some eighty-three thousand years back, stood out as a major contributor. This is an exciting twist in the story that will have us rethinking apple history and genetics.

Subsequent studies analyzed the nuclear genomes and revealed that, while much of the apple's nuclear genome came from a *Malus sieversii* ancestor, a significant portion of the apple genome was acquired from *Malus sylvestris*, and another portion from both *sieversii* and *sylvestris*. Roughly 30 percent of the apple genome is heterozygous and of hybrid ancestry. The lack of reproductive barriers between apples and crabapples, together with the self-incompatibility of the apple tree, allowed genes from the European crabapple to mix with the cultivated apple.

While generations of scientists debated the apple's origins, it has turned out to be a classic hybrid. Though it initially came from central Asia, during its travels along the Silk Route, it hybridized robustly with the European crabapple. Behold the antics of a world traveler.

II

We selected the sites for our crabapple and three apple trees carefully, after observing how the soil drains in each site, especially since our clay soils can be very slow to drain, and roots can rot in standing water. Through a microscope, clay particles look like tawny, flat plates or tiny flakes. The pore spaces of clay particles are too small to allow for easy movement of water or entry of air, which is why clay soils must be amended with organic matter in order to grow trees in them.

When planting the trees, we avoided the paths the deer favor when they walk through our patch of land. Garrett had told us that at Tooley's orchard, their browsing problem isn't deer (an electric fence deters them) but coyotes who jump up and get at the apples or break a branch. In Indigenous lore, the coyote is said to have instructed humans about which plants are edible; no doubt the coyote wants her rightful share of our fruit. Coyotes do wander through our land and howl almost nightly along our dirt road. Our surviving fruit trees, however, contend mostly with browsing deer and, on occasion, rabbits. One of our apricot trees sprouted from seed, but rabbits munched off all its tender leaves.

Our apple trees will no doubt need to be fenced, but we face another formidable problem: Our clayey soil compacts and hardens when dry, and it is hurting from the drought that has persisted for twenty-two years. The American Meteorological Society defines drought as "a period of abnormally dry weather sufficiently long enough to cause a serious hydrological imbalance." While drought is not a new event in New Mexico, climate scientists have noted that this "recent drought has been unusually severe relative to the droughts of the last century, but some droughts in the paleo climate record were much more severe." With due respect to paleo climate records, today's young fruit trees crave water. A recent analysis led by a University of California, Los Angeles climate scientist confirmed that the West is experiencing its worst drought in 1,200 years.

The prognosis for the Southwest's water resources is grim, but in cities like Santa Fe, not only do numberless cramped, ill-conceived developments go on rearing their heads, erasing wild spaces, but the city also continues stretching to the north and south and, heedless of wildfires, pushing up against its mountains and deserts. If our politicians are ruminating over the existential crises brought on by climate change and the megadrought, they have yet to offer proof.

A foundational tenet in geology is that the earth is a closed system, and the amount of water on earth is constant. Seventy-three percent of the surface of the earth, what the poet Jim Harrison called "the watery face of the earth," is covered in oceans. But only 2.5 percent of water on earth is freshwater; two-thirds of this freshwater is in the form of glaciers and ice caps, and nearly a third is groundwater. Only a staggeringly low 1.2 percent is the liquid surface water that we see in lakes, streams, and rivers. It is this evanescent sliver (and the subsurface aquifers) that, essentially, cities and towns deliberate over—should it be diverted to, say, fruit orchards or golf courses? What we do with our limited usable water is a foundational matter that says everything about who we are. An old Spanish saying pervades New Mexico farming culture: "Agua es vida." Water is life. One farmer, Nicholas Herrerra, adds: "Without water, there is nothing."

We aren't alone in worrying about the impact of drought on our small orchard. Local farmers tell me that their young apple, pear, peach, and cherry trees have simply died over the last few years. Even farms with three-hundred-year-old water rights are struggling with not having enough water for their fruit trees. On the Green Tractor Farm in nearby La Cienega (the village the Real Orchard is in), three years ago a young couple put in an orchard of some thirty apple and pear trees and one peach tree. The young trees were planted in the fall; the acequia, or irrigation ditch, is turned off in the winter and the trees weren't watered then. The soil had grown so dehydrated, however, that the next year fully a third of the trees gave up the ghost.

It isn't just fruit trees that are signaling that our faltering water supply, exacerbated by climate challenges, will haunt us sooner than we realize. A state report on forest health in New Mexico stressed that dehydrated trees of all types are unable to develop healthy canopies or create the sap they need to repel insects; in our forests, piñon and ponderosa pines, and mixed conifers, are dying of thirst across tens of thousands of acres. "Climate change is here and now," Senator Tom Udall writes, "and, in the arid Southwest, New Mexico is right in the bull's eye—with temperatures rising, drought on our doorstep, snowpack and spring runoff anemic, and wildfires threatening."

The apple tree is adapted to a temperate climate and must experience a set of narrow temperature conditions in a certain sequence throughout the year in order to produce quality and abundant fruit. As the seasons change, an apple tree undergoes physiological changes. Autumn brings fewer daylight hours and cooler temperatures, signaling the tree to prepare for winter. In this period, it develops terminal buds, halts growth, and begins the "hardening" process that will enable it to survive subfreezing temperatures. Water is supercooled inside the cell tissues or is transported out into the intercellular space, where it can freeze without causing damage to the cells. As cooler temperatures intensify, the tree enters dormancy. For proper development of flowers (and eventually fruits), the tree crucially needs to accumulate a certain number of chilling hours, typically three to four weeks at temperatures from minus 2 degrees Celsius to 13 degrees Celsius, to complete its dormancy cycle. The chill-hour requirement makes apple trees particularly susceptible to climate change, as the number of winter chill hours in apple-growing areas grows smaller. When warm winter temperatures cause apple trees to break their dormancy too early, subsequent frosts can be brutally damaging.

A paper in the journal *Environmental Research Letters* lays out six metrics that can adversely impact apple growth. Yakima, the largest apple-producing county in the US, which produces well-known varieties including Cosmic Crisp, Honeycrisp, Fuji, and Gala, has already experienced significant changes in five out of these six metrics: increased exposure to extreme heat (leads to sunburn and heat stress); more warm summer nights (affects coloration); the number of cold days; fewer chill hours; earlier last day of frost (impacts timing of apple blooms); and the number of days above a certain temperature that are needed for apple growth. After studying forty years of climate data, Washington State University researchers determined that the Yakima orchards have already been harmed by the changing climate.

The RAFT Alliance cautions that within three decades we will lose apple production in key areas such as the Central Valley of California and "from southern Pennsylvania, as well as from many warmer localities found at lower elevations across the continent." The nation's apple experts exhort us with hoarsening voices to work together and "ensure that apple diversity will contribute to our food security in the future, and not disappear in a time of rapid climate change."

III

Healthy soil is the mother of land and life. But what is the relationship between land and soil? Aldo Leopold believed that land is not merely soil: "It is a fountain of energy flowing through a circuit of soils, plants, and animals.... When a change occurs in one part of the circuit, many other parts must adjust themselves to it." Global warming has introduced huge changes into the circuit—not remote or future changes, but ones we are reckoning with: plants and trees blooming too early or wilting in the drying soil; birds and other animals growing scarce. Reflecting on the scrawny coyotes who now crisscross our land, and the tanagers, kingbirds, and orioles who rarely come anymore, I suspect that they are also having trouble eking a living from this land.

The dry winters have been especially hard on the coyotes. One March evening, we were at the dinner table when a coyote walked by. She was lean, too lean. We congregated at a window; the coyote looked us in the eye before crossing the backyard and disappearing among the piñon pines on our western slope. A rabbit downwind sensed something, for he spun around and raced off. The coyote crossed the dirt road and walked along a narrow arroyo on the other side. A little later, we saw her wearily retrace her steps along the east side of our house. Apparently, she hadn't found anything. Michael observed that the way she was walking, she was too slow to catch a small animal. The too-lean coyote is a living example of how less energy is flowing through the circuit.

The soil grows the plants the rabbits feed on before themselves becoming food for the coyote. But why is the fertility of our soils waning? Leopold defined fertility as the ability of the soil to receive, store, and release energy. In its native state, soil has a rich capacity to hold water and nutrients. But industrial pollution, greater evapotranspiration, and pervasive bare soil, along with lower density of native plants, have reduced the carrying capacity of our soils.

"Soils depleted of their storage, or the organic matter which anchors it, wash away faster than they form," Leopold wrote. His prophetic concern about soil erosion in the Southwest has been borne out.

Weakened by drought, junipers and piñon pines in New Mexico grew more vulnerable to bark beetles, resulting in the deaths of some 350 million piñon pines—a keystone species that holds together the soil in our fragile ecosystem—and aggravating erosion. It is remarkable that the piñon pines are weakened by this drought given that their tap roots can extend more than forty feet deep in search of water. Over the last fifteen years, I have observed the steady erosion, too, of our slopes and those in neighboring areas, all of which have grown steeper; roots of young conifers now stick out in the air when they were once firmly entrenched in the ground. We began composting not long after we moved here. But when the soil is bare and bone-dry, it capitulates to short, hard rains that run off and wear down the land. What measures can we take to improve the health and stability of this doubly beat-up soil? How can I dream of growing fruit trees here so that my children, and many others, will intimately know the apples and apricots that once thrived here?

In our garden, after over a decade of mitigation, our native perennials are doing okay, but we still can't seem to grow fruit trees. The fruit trees we bought ten years ago at Tooley's farm have struggled to survive. We made a real effort to care for the soil and the newly planted trees—fruit trees do need more water, and we supplemented their drip irrigation with recycled kitchen water. Still, the peach and the nectarine trees died. The apricot trees are still no more than four feet high, and the hawthorns are no taller than three feet. I was pregnant with Mia when we first put Tooley's trees into the ground; she has thankfully grown at a faster pace than the trees.

A decade later, we are anxious to do better with the apple trees. We'll need to keep a closer eye and tend them as though they were toddlers, while being mindful of the drought—they'll need more compost to compensate for the low decomposition of organic matter in a desert ecosystem, and at least four inches of organic mulch to reduce evaporation. If we can cultivate healthy soil, it may also deepen the flavor complexity and terroir of the apples.

One evening, I telephoned an old neighbor from Haflong, Assam, and told her about my quest to grow apple trees.

"Well, if you plant them now [in the fall]," she said approvingly, "you'll be able to eat the fruit next year."

Clearly, she has never been to New Mexico.

Chapter Eight

Pulling the Crank

OCTOBER

Apples at the Real Orchard. Photo by Priyanka Kumar.

I

A few days later, Michael and I took a field trip with Gordon Tooley to the Real, the oldest orchard in our area. I wanted to untangle the existential challenges that a historic orchard faces and understand how this parcel of land had so far resisted fragmentation and had instead entered the ranks of the micro-wild. A flock of chattering black-billed magpies soared over us, flashing inky blue, black, and white, as we drove up to an acequia that borders the Real Orchard in the village of La Cieneguilla, forty minutes south of Santa Fe. The acequia or irrigation ditch ran as far as the eye could roam. By the dirt road, a small sign read, "Real Ditch, Circa 1718." We were entering a 1,200-acre property on a small part of which stands an old orchard, but the sign caused me to look again at the acequia that has been feeding thirsty trees here—since 1718.

A ditch can represent life itself—so, it may be cavalier to call an acequia a ditch. The word *acequia* comes from the Arabic *al-sāqiya*, which means "water conduit." An acquaintance who'd spent the summer in his homeland in North Africa told me, "My father has an *al-sāqiya* by his house even today." The Moors honed their survival techniques in North African deserts before occupying Spain for nearly eight hundred years, until 1492. So, it was the Moors who introduced water management methods to the Iberian Peninsula. The Spanish colonialists in turn brought these irrigation techniques to what is now the Southwestern US, diverting water from a river to an *acequia madre* or mother ditch, which often runs along the top of the slope edging a field.

On one side of the Real ditch, spiny cholla and stunted junipers eke out an existence on arid land; on the other side, a line of native mulberry trees with yellow-gold haloes, roots hewing the ditch, edge the vast field where Alonzo Real once grew vegetables. Now the field is scratchy with weeds, but the fruit trees on the southeast side of the property are miraculously thriving. The acequia accomplishes that. It gives us the shade and beauty of the row of fruitful apple trees and one quince tree that skirt the far end of this field. This so-called ditch has been a giver of life to Spanish farmers since 1718, in the same way that the Santa Fe River flowing nearby in dense green rivulets was a source of life to the Keres Puebloans whose lives were much disrupted when the Spanish arrived.

Michael was driving and Gordon Tooley rode in our back seat; he jumped out to open a metal gate. "I hope the code I have still works," he said. "The BLM has a way of changing it." Fortunately, the code worked, and we drove up to a Spanish-style hacienda built around 1890. On the east side of the pink-stucco house towered two trees of heaven, *Ailanthus altissima*. "Those trees are grandfathered in," Gordon said with a gloomy shrug, about the invasive species.

Taking in a sweeping view of the land, I was struck by its splendid brooding air; the old house and an adjacent field seemed to be cut from the same ancient cloth. Who had lived here and how did this land survive the ravages of the twentieth century? Primeval basalt hills loomed over us, framing a wide field edged on its south side by mature fruit trees. This ribbon of hills stretches all the way to Socorro, a two-hour-drive south, and the hills are rich in petroglyphs from a time when Native Keresan peoples lived here in pueblos until the early 1600s. Still older petroglyphs have been dated as being eight thousand years old. The land sings with history.

"We find pot sherds everywhere," said Keer, who lives in the only other house in the vicinity. A white-haired, white-mustached man of Irish stock, he had just joined our party.

The rain picked up, graying the air. I told Keer that he was fortunate to live among remnants of the lives of the Keresan peoples.

"Things were fine back then, in 1491," Gordon added with a smile. Meaning before the Europeans came with their exploitative and irreverent attitudes toward the land.

The pre-1821 Spanish Land Grant distributed a staggering three million acres from the king of Spain to Spanish settlers in New Mexico—which is how the Real family ended up with 1,200 acres in the village of La Cieneguilla. In 1917, Alonzo Real was born in this hacienda; the house and land surrounding it would be his home for the next eighty-six years. In the twentieth century, a restless era marked by migrations from village to town to city, and from country to country,

I wondered how and why Alonzo had remained rooted to this land for so long.

As a young boy, Alonzo saw vegetables and fruit being grown on the land. "His family would take the vegetables to Santa Fe," Keer said. "They'd sell them near the railroad track."

I nodded. That is where the farmers' market stands today.

"By wagon, it would take two days to get there and back."

"That includes some time to get drunk," Gordon added, chuckling.

Keer moved next door twenty years ago, and the field we stood on flourished then with alfalfa and chile. Back then, Alonzo was still growing vegetables in a large garden by a Siberian elm at the far end of the field.

Today, the five-acre field lies fallow, and Gordon hopes to keep it from becoming "a weedy and woody patch." Next spring, he will plant some annual seeds to improve the soil and move the field a step closer to becoming a perennial system, rife with native grasses. Across the field he pointed out "Solo," an apple tree with dramatic bones, in conversation with the sky—the tree is on its way out. It's too late for Solo—Gordon doesn't think that his interventions can help. He agreed to become involved with the Real Orchard only after some experts "butchered" the fruit trees, which is Gordon's way of saying that they pruned them incorrectly.

The first time that Keer and his wife Carol saw this field, an Easter egg hunt was underway for Alonzo's extended family. Intrigued by the striking landscape and a white cross painted on top of a neighboring hill, Keer and Carol were driving around when they came upon the egg hunt. But they felt conscious about being "Anglos" and didn't walk in on the festivities of the large Hispanic family. Still, the land tugged at them. Two years later, responding to a newspaper advertisement, the couple purchased a three-acre parcel next to Alonzo's—which had been his nephew's home. They soon solved the mystery of the white cross.

"Alonzo painted it to give thanks for everyone who went to World War II from this village and came back alive," Keer said. His eyes lit up when he spoke of Alonzo, and I wondered how the two men became friends. We had been standing out in the persistent rain for some time, and Keer suggested that we wait out the rain in Alonzo's pink-stucco house.

II

In the milk-white foyer, some of Alonzo's books still stood in a built-in bookshelf. One on folk medicine, an encyclopedia of unusual diseases, and an illustrated book on Latin America. A bill for coal purchased over two decades back lay on a lower shelf. To the left of the foyer was a light-filled bedroom Alonzo had used except when he migrated in the winter to the back of the house, to a bedroom with a heater.

Keer remembers Alonzo as a sweet-tempered man who was never judgmental. He didn't marry but enjoyed spending time with extended family. His house had no electricity or running water. Once, when there was an opportunity to get electricity, he said, "What do I need that for?"

"What do I need that for?" is a fitting question before making any new acquisition. (I recently laughed out loud when Greg Ohlsen, the proprietor of a local store, Travel Bug, said in an interview: "Most people buy more than ten times what they need.") In a recent state of the climate report, a team of scientists led by William J. Ripple called climate change "a glaring symptom of a deeper systemic issue: ecological overshoot, where human consumption outpaces the Earth's ability to regenerate."

As I took in the interiors of Alonzo's home with its whitewashed walls and the windows opening out to fruit trees and hills, I experienced a familiar insight: When the universe filters into your living room, your needs tend to grow sparse.

The back bedroom led to a kitchen where Alonzo's electric-blue wood-gas-and-coal stove was still standing. "He was into herbs and homeopathy, and he dried and processed his herbs here," Keer said, a nostalgic look charging his eyes as he pointed to a corner sink.

I was beginning to understand why the land has retained its ancient character.

"When Alonzo was in the house, he spent 95 percent of his time in the kitchen. . . . I've spent many hours sitting with him at this table."

A glow came over Keer's face as he recalled those hours. Alonzo may not have needed electricity, but the air in the kitchen turned electric as we spoke about him. A generous soul, he had brought light and joy into his neighbor's life. A contented silence hung in the air, and all four of us felt as though we had shared something essential. I know little about Alonzo, but I know that one marker of a meaningful life is that you continue to stitch people together years after you are gone. I know this because of people who have materialized out of the blue to tell me how my parents touched their lives.

Alonzo's sister, a resident of this village, lived to be a hundred, and passed away five years ago. Once a week, Alonzo went to fill five water drums of ten-gallon capacity at his sister's. "He stored his water in the kitchen . . . and shaved in here." Keer pointed to where a mirror hung next to a door that led outside to the trees of heaven.

The back door led to an open corridor, which faced a gnarled cottonwood, and a grassy path to an outhouse I will avoid using. Next to the outhouse, Gordon once encountered a six-foot-long rattlesnake. Keer topped that by finding an almost seven-foot rattler on his land next door. Rattlers roam freely here. Once, when Gordon drove in through the gate, he all but ran one over. He got out of the car to look at the snake and for some reason poked a stick at it; the rattlesnake hissed furiously.

III

It was still raining, but we returned to the field. Taking in the luminous wet beauty of the orchard, I thought aloud that Thoreau might have liked to plant a hut right about here.

Keer chuckled. "And the pond is right over there. . . . This is New Mexico's Walden."

The comparison struck me with force. While Thoreau had lived in Walden for roughly a year, Alonzo lived in his Walden for eighty-six years. He is a fortunate man who is born into a landscape and goes on to love it for a lifetime.

Keer left soon after, and Gordon, Michael, and I hiked on to the pond. Like a large emerald hidden by weeds, the pond didn't reveal itself until I was nearly at its edge. Three-quarters of the surface was layered with avocado-green algae. The resident heron was absent, but a midsize bird, a lark perhaps, fluttered out of a tree and fled, all but grazing my face. The rain intensified as we bushwhacked through the slushy ground; my sneakers were soaked through, dozens of burrs clung to my leggings, and my thin woolen cover-up was barnacled with thorns.

Crossing the field, we made our way to the southeastern apple trees. The acequia ran right alongside the trees, suckling them like a mother. After wading through thick weedy vegetation and hopping over ditches flowing with muddy water, it was glorious to come upon a row of eight apple trees laden with deliciously sweet and tartly crisp fruit. No bear or deer had been here lately for the ground was thickly littered with apples. Most likely these were deep-flavored Winesaps and sprightly Jonathans—later I would learn that the Wolf River variety also grows on this property. To identify the varieties with more certainty, Gordon quartered the apples, taking pictures of the "star" inside and of their tops and bottoms (calyx). Besides these markers, the water core and flesh are additional clues, together forming the "fingerprint" of an apple.

"And if I don't find out what they are, that's fine too," Gordon said. "There are things I don't know."

I nodded. The land has its language, and we try to read it insofar as we can. Six-year-old Pika is trying to read by herself. Last night and this morning, she was reading *Click, Clack, Moo*, a book in which cows know how to type, struggling with a few unfamiliar words, like "impatient" and "impossible." So, there are varieties that Gordon can't identify by sight, and there are things I am still learning—but I want to piece together the poetry of this land.

The mulberry trees edging the northern end of the field, and yellowing radiantly in the fall, tell their own story. Back when the Department of Agriculture (established in the 1860s under Abraham Lincoln) was asking everyone to plant Russian olive trees, Alonzo followed this advice but soon discovered that he hated the trees. Swearing vociferously, he tore them all out. Instead, he planted the native mulberries that flare brightly in the sun, tugging the heart as they do in Van Gogh's painting of a mulberry tree in the last days of his life. In Benjamin

Franklin's later years, he too planted mulberry trees prominently in the modest garden of his Pennsylvania townhouse. *Morus celtidifolia*, Texas mulberry, is one of the few native fruit trees in America; introduced to the Grand Canyon area by the Havasupai people, its savory-sweet berries paint the tongue purple.

"We didn't get to them this year," Gordon said. "But last year we put a tarp underneath and shook the trees. We got a lot of mulberries. This year the birds got them."

I grinned, feeling delighted with this outcome.

On one apple tree, a large nest constructed with coarse twigs was nestled on a forked branch. It housed either a squirrel or a magpie. In the ditch, the water flowed steely gray, and sage-green grass and weedy vegetation grew beyond. Steering around burr-filled plants and beavertail cacti, I eyed the blushing yellow apples in the branches above and experienced abundance, a feeling of being surrounded by riches. I hadn't brought water along on the hike, anticipating that I could reach for an apple if need be. Apples quench my thirst better. These apples are life-giving. The land throbs with life and, in the silence, I hear the unfamiliar hum of all that transpires in the earth, which is mostly invisible to the eyes. This may simply be a row of apple trees, but as I wove through the trees under the gaze of ancient hills, I experienced an intimacy with the trees—and an awakening that this is where I want to be. To be clear, I felt at home.

We surveyed and took samples from five of the seven apple trees that were abundant producers, and I didn't take anything from the quince that produced less fruit. The quince stood proudly like Snow White amid Seven Dwarfs, silhouetted against the basalt-topped mesas and the azure sky. Gordon's face lit up when he realized that the quince tree was bearing fruit. He stretched up to pick off two quinces, then jumped smack into the tree fork. Climbing up, he lowered a branch so that Michael could pick off a third quince.

"I'm bad," he said.

I demurred. Picking fruit is an intrinsic part of our engagement with these trees. While it has been lost in our everyday lives, it can still be restored. A close relative of the apple, the quince cannot be eaten straightaway—it needs to be processed first. "It has fuzz on its skin and has to be fermented completely and made into a paste before it can be eaten," Gordon said.

As we walked on, he pointed out the diseases or problems some of the apple trees had suffered, such as black rot and sunscald. We forged a path through tall weeds and could scarcely see the ground beneath. Gordon wandered insouciantly, used as he is to rattlesnakes. My own fascination with snakes is mixed with wariness after a host of encounters and stories of a bite that took down an experienced herpetologist and another that surprised a veteran wrangler. Gordon grew up in the town of Cimarron, east of Taos, and likely dodged some snakes while wandering through old homesteads to gather apples. He was also at the Real over the spring and summer—prime rattlesnake season—to water and prune the apple trees to stimulate healthy canopy growth.

On how not to prune a tree, he said: "When you cut off the top canopy, the tree responds by doing more vertical growth, in a bid for survival." This type of growth stresses the tree in addition to making it more vulnerable to sunscald. I noted how the upper branches of an apple tree I stood under had been scalded by the summer sun. Gordon favors a rounded-out canopy, which is healthier, and he looked satisfied with how the trees had grown since he'd last pruned them. Under his care, all the surviving trees now bear abundant apples.

IV

We walked back in the direction of the old house, which blended with the landscape, and continued on beyond. Sauntering through the land, I felt as Hermann Hesse might have after he left behind his middle-class life and walked through the countryside for days on end. "When I see this blessed countryside again, at the southern foothills of the Alps, then I always feel as if I were coming home from banishment, as if I were once again on the right side of the mountains," he wrote. "Here the sun shines more intimately, the mountains flow with a deeper red, here chestnuts and grapes, almonds and figs are growing, the human beings are good, civilized, and friendly, even though they are poor. And everything they fashion seems so good, so precise, and so friendly, as if it were grown by nature itself."

This is how Alonzo's home looks, as though it is good, precise, and friendly—and perhaps grown by nature herself. The fields and landscape beyond seem to emanate from the pink stucco house and then fold back into it. I found myself daydreaming about living in the house, at least from spring to fall. Then I recalled the outhouse with its resident rattlesnake.

Gordon paused to show off a tree and asked me to guess which tree this was. He smiled when I guessed correctly: "Meat Hook," the pear tree with hooks to hang a hog from. Up ahead, an apple tree's base was clear of fruit and the grass flattened as though by deer. Gordon thought that the greenish-yellow apples on this tree must be very tasty since deer had devoured the windblown fruit. But when we tasted the apples, they were slightly mushy and on the verge of being bland. Perhaps the deer relished the bland flavor? Our shoe soles slick with mud, we walked on to the west side of the property, crossing the Santa Fe River on a single plank bridge that was soaked through. Gordon had installed a water pump here, atop an incline. When he climbed up to it, he muttered and swore. "Someone has cut my lines."

The cuts were clean, so an animal hadn't made them. Anyone can hop the gate and wander through the property. But why cut Gordon's drip lines? Why disrupt efforts to heal this all but dying orchard? We hiked up to a solar panel that he uses to power the water pump, which was still intact. "I'm surprised it hasn't been stolen," he said. Tucked into the solar panel was a nest. "I saw the eggs and the chicks.... They'll be back next year."

We walked over to the rest of the fruit trees, including a couple of apple trees, and plum and pear trees. The two apple trees that survive are of the Winter Banana variety, so called because the apple—tart with mellow-sweet notes—ripens to a lovely autumnal yellow; the Winter Banana originated in Indiana and is a great pollinator for other apple varieties. The other fruit trees were blown down during the storm; it didn't help that the roots had been previously gnawed at and weakened by gophers—a sign of neglect. This row of fruit trees borders a second field, where Alonzo grew vegetables. Gordon pointed out that the scrub here is stunted. So, Alonzo would have had to finely read the land here in order to irrigate correctly.

"He might have done some dry-land farming here," Gordon said, referring to a method that relies on nature, not irrigation, to water crops—and which is gaining traction once again in a water-starved state.

It was in this area that the irrigation ditch busted after a climate-gone-awry storm and wasn't repaired (after Alonzo's death), and eventually led to the death of the mature apple trees. Acequias have to be cleaned with diligence so that water can move through unimpeded, but the culture of maintaining acequias is also fading. A few months later, I would come upon a *mayordomo* in Taos who was supervising the annual cleanup of his local acequia. "I have five hundred *parcientes* (members) in our Acequia Association," he told me. "And guess how many showed up to clean it today?"

"A dozen?"

"One." Forced to hire youngsters to do the work, the *mayordomo* lamented that he was kept busy dealing with their flare-ups and absenteeism.

At the Real Orchard, I gazed at the skeletons of the apple trees laid out on the ground—bleached and dreary remains of what were once flourishing beings that could nourish themselves and us. If Gordon is able to pull together some resources, he will have the tree rings dated. He suspects that these trees may be 120 years old. If these trees could talk, I wonder what tale they would tell about perishing from thirst.

Gordon told us that there are a few haciendas like this one scattered across the state and on Hopi and Navajo land. Sometimes the will is there to take care of such places, but it can be short-lived. When I asked about BLM's vision for this land, Gordon said anxiously that there is no saying what will happen. After Alonzo died in October 2004, his brother-in-law was appointed the executor of the estate, and he later put the Real property on the market with an asking price of between $2.5 and 3 million. In 2008, the economy crashed and there were no takers. The property was subsequently offered to the BLM for $1.4 million. Gordon and others hope that the structure will become a historic museum and residence for the caretaker of the land. The BLM can be an unpredictable agency, interested in monetizing land, and Gordon and Keer's lone contact there, a sympathetic engineer, plans to retire in a year.

We returned to the pink stucco house and sat on rocks outside to eat our sandwiches. An intimacy came from eating together in the open and sharing apples; if Alonzo were watching from the kitchen window, he might have approved. Soon, Gordon was speaking his mind and asking the question that will determine this land's future: "Who will pull the crank?"

I nodded, recalling that some machines require an individual to physically turn a crank—an essential lever—to operate them. Who will put in the manual effort it takes to work this land? The caretakers of the land seem to have gone missing, which is one reason why similar sites do not survive. The Real is squeaking by, with Keer doing some voluntary caretaking. When Keer saw some trees drying up, including a line of *Robinia pseudoacacia*, black locusts, that lead up to the house, he steered water from the acequia, diverting it to a slender ditch that he ran by the trees to bring them back to life.

"They're still a bit dry," Gordon said. "But now they'll be fine." If the black locusts were to go, that would mean the land is really stressed. The locusts spoke to Keer, and he responded. But will the BLM respond when this land deteriorates further?

"The BLM owns a lot of land around this property, which is one reason they snapped up this parcel," Gordon said. "For the BLM, this property is just a postage stamp in their collection."

This land, steeped in history, is nothing short of a treasure. But it has become harder to be faithful to the land as Alonzo once was. Even after thirty years of running the tree nursery, Gordon doesn't have any solid apprentices. Talented interns go through, but it is a hard way to make a living; in the end, one might be swayed by city living, another by weed and the internet. A hyper-digital world offers us a dizzying array of choices, and it's perhaps not surprising that we have lost our sense memory of the land. But it lies buried in our souls. Gordon and his wife don't have children, and he doesn't know what the future of his farm will be. He confessed that over the last five years, his farm has been on "life support," and the contract work he does to restore other people's land is what keeps his own farm operational.

A flicker cried *kew kew* before flying away from the canopy of an elm. The land murmurs and, when we listen, it sings old hymns—birdsong is only the beginning. Listening to the land feels like an urgent skill at a time when so much of it is being erased and degraded. There are still Waldens among us, though one has to cast a wider net to find them. Lost in the maze of our devices, however, we may collectively lose our facility to sense nature's rhythms. If we cannot perceive her rhythms, how will we recognize the changes that global warming has wrought?

Gary Snyder points out that land is not a "fixed text." I agree, but that is no reason to discard the metaphor of land as text. Land is a dynamic text—and if, in a digital era, we can grow conversant with new forms of storytelling, surely we can relearn the language of the land. If a six-year-old can learn to read a story about cows who type, surely we can absorb the grammar of trees, soil, sky, and birds. If animals could really type, I wonder what they would say to us.

Chapter Nine

Wild Horses

NOVEMBER

Duchess of Oldenburg apple tree at the Rayado Orchard.

I

The Real Orchard and the landscape surrounding it percolated into me, and I felt stirred to wander in more historic orchards—where mature heirlooms grow and blend into the landscape. The Real was one person's twentieth-century passion, but were there older orchards left that had once been planned on a larger scale? Tooley told me about his friend Dave Kenneke, a ranch manager and apple lover in Cimarron, New Mexico, and gave me his number. At the end of October—peak apple season—I called Dave and left a message. Days ticked by as I waited. Gordon's question rang inside me: Who will pull the crank? After so much land literacy has been lost, the few caretakers who are left are stretched thin.

A week passed, another began, and I gave up hope of hearing back from Dave. One incandescent afternoon, the phone rang. A hearty but distant-sounding voice spoke to me from a mountaintop. Dave said that he had been traveling. "I'm extraordinarily tied up," he added.

I told him that I was researching historic orchards and was a pilgrim on the apple trail. All at once, Dave sounded refreshed, and we went deep into conversation. It ended with him suggesting an overnight stay in Cimarron so that I might experience the Chase Orchard in Colfax County. I hung up, feeling exhilarated. I wanted to bring my family along, and Dave had offered to dig up some dorm-like accommodation. Little did I realize that we would soon find ourselves walking 150 years into the past. But the past isn't entirely past while some historic fruit trees survive. The somewhat distinct apple phases—settler and Spanish orchards, apple evangelists, the golden age, and industrialization—also blend into one another. For instance, the apples the Spanish introduced persist even today in the Manzano Mountains and in Rio Arriba County. But they have intermixed and hybridized with nineteenth-century apples from the East and Midwest. When I am studying a historic orchard, it feels like I am walking through the pages of a history book. In the third phase of America's apple love, the nineteenth century's golden age, many farmers in the East "converted their fields from cereal grains and vegetables to orchards" and, in the West, "many newly claimed lands were immediately developed as commercial orchards."

One early November afternoon I forged through thick weeds to a dirt path that runs through the Chase Orchard. Few orchardists better exemplify the nineteenth-century story of the "quintessential homestead fruit" than young Manley Chase. In Colfax County, Manley Chase and his wife, Teresa—transplants from Colorado—set up one of the largest orchards in the area and grew apples that won national renown. A thrill went through me as I wove through apple trees that were planted in the late 1800s and still bear fruit today. Apple trees have an average lifespan of 80 to 140 years, and many heirloom trees are now nearing the end of their lifespans. Which makes it urgent to know magical 140-year-old beings and propagate rare varieties before they are lost to us. In the early 1800s, American nurserymen offered some "100 named varieties for sale; by 1850, more than 500 widely recognized varieties were being cultivated." In 1872, around the time when Manley Chase began to dream of apples, a national inventory was made of "close to 1,100 different kinds of apples that had originated here in America"—this is considered to be the first estimate of apple diversity in the country.

In 1871, aiming to ease up on rearing sheep and growing oats and cover crops such as alfalfa on his new ranch in Cimarron, Manley ordered some apple, plum, and pear trees to be shipped to him across the Old Santa Fe Trail. Soon he was planting an orchard south of where he and Teresa planned to build their new home. Manley hired men to dig irrigation ditches, the first of many that he would set up along the Poñil Creek. The fruit trees did well that first year, proving what he had suspected—and he felt brave enough to order 250 more fruit trees from Ohio.

"One day the wagon loaded with Manley's new fruit trees drew up to the freight station in Cimarron. It seemed the whole town was there to watch." Locals ridiculed these early attempts at growing fruit as "Chase's folly." In cattle country, it was seen as insanity to devote fifty acres to fruit trees. Manley, however, turned out to be no fool; his correspondence with the secretary of agriculture in Washington, DC, and the agricultural tomes he purchased can be found in his pink adobe house even today. Manley and Teresa would thrive in this beautiful, fertile valley, and their fruit trees would mature into a spectacular orchard. Their story exemplifies how American farmers prospered in the golden age of pomology. At its peak, the Chase Orchard had six thousand fruit trees and produced a staggering five hundred thousand pounds of fruit each year. At the Chicago World's Fair in 1893, Manley's apples won a gold medal. In a US congressional publication from the same year, Chase Ranch was listed as a leading orchard, and the assessment was confirmed in a 1901 report from the governor of New Mexico to the US secretary of the interior: "Fruit growing and the management of the crop have been brought to a business standard on these big fruit ranches that is not excelled anywhere in the West, and certainly not in the Rocky Mountain region."

In the 1920s, New Mexico state publications still touted that Colfax County apples led the world in flavor and quality. No one keeps count any longer, but hundreds of apple trees still brushstroke the land as far as the eye can reach. This forestlike aesthetic won me over, but I also wondered why the generations that followed didn't tend this magnificent orchard. Why did it wear an abandoned air like so many orchards I have experienced?

At the Chase, the apple trees from the 1870s had been waning until recently, when a man named Ernesto, who scarcely knows English but is fluent in the language of trees, came along and pruned them so skillfully that the trees once again grew bountiful. But Ernesto, Thelma Colker (a master gardener who tends to the Chase Orchard), and Keer (the voluntary caretaker of the historic Real Orchard) are all in their late seventies and eighties and perhaps ready to put down their shovels. Thelma made a curious observation. She told me that Chase's son, Stanley Jr., had also loved apples but didn't have the "energy" to do much for the trees. It is true that caring for the land can be punishingly hard work. But as I would soon learn, tending established apple orchards with heirloom varieties, in the midst of community, can also be surprisingly joyful. Regardless, in the twentieth century, our collective energy for apple orchards seemingly flagged. According to government records, "approximately 50 percent of the fruit trees that existed in 1880 were gone by 1930." Why did this happen?

Cimarron means "wild" in Spanish, and the striking orchard the Chases put in is girdled by mountains on all sides and retains a flavor of wildness. Not long back, Gordon Tooley planted some young trees here, grafting old Chase varieties that had survived onto healthy new rootstock.

"He's one of my heroes," said Dave Kenneke, a tall, robust man who carries himself lightly. The director of ranching and conservation at the nearby Philmont Ranch, Dave, along with gardener Thelma Colker, walked me through the Chase Ranch on the late autumn afternoon. All the while, Mia and Pika climbed every-which-way on the fences that ran along a cow pen and raced around the cowshed, delighting when Dave's dog, Rusty, gave in to ancient instinct to lord over the cows.

Manley Chase wanted to grow hardy varieties of apples that would survive a train ride to the East Coast and a stay in people's cellars in an era before refrigeration—and remain fresh until December. "Back then it was thrilling to have fresh produce in the winter," Dave said. A week back, he had dug up for me an undated typed list, titled, "Varieties of Apples in Chase Orchard." Forty-nine varieties are listed, including five varieties of crabapples. Among the latter is a Winter White, which Thelma was likely referring to when she said, "There's a white crabapple around the corner that makes such good jelly!" Her kind, crinkled face glowed in the last flare of afternoon light before the orchard's cool shadow drenched us. Other listed apple varieties include Black Dragon, Winter Banana, Strawberry Spice, Wolf River, Sheep's Nose, Arkansas Black, Maiden Blush, and Nonesuch. What a faded world in which our children have never heard of, much less tasted, such enthralling varieties. These varieties may sound exotic today, but they were once dismissed by Thoreau, who favored wilder apples over grafted ones: "Indeed, I have no faith in the selected lists of pomological gentlemen. Their 'Favorites' and 'None-suches' and 'Seek-no-farthers,' when I have fruited them, commonly turn out very tame and forgettable. They are eaten with comparatively little zest and have no real *tang* or *smack* to them." Thoreau and his brother, John, fervently planted apple trees and a melon patch and were known in Concord for a summer party they threw when the melons were at peak ripeness.

On the two-and-a-half-hour drive north to Chase Ranch, we had stopped at a café in the town of Watrous. As I walked up to the café, I saw a mature apple tree growing in the center of an adjoining lawn. The tree grew wild, or at least it looked unruly and untended. I dashed over and picked its last viable apple of the season; it was certainly tart and couldn't be accused of lacking *tang* or *smack*. It gladdened me that someone had let the old tree remain and perform its wild magic. Later, I would learn that a Sheep's Nose apple tree in this town is one of the largest in the state, with a circumference of 152 inches and a crown spread of thirteen feet. The elongated shape of the apple is reminiscent of a sheep's nose, and the fruit has a whiff of cloves; this old American variety originated in Connecticut in the early 1700s. The Sheep's Nose is one of five remaining apple trees in a private orchard garden in Watrous; planted in the early 1860s, the trees supplied fruit to travelers along the Old Santa Fe Trail.

II

Manley and Teresa first moved to this broad valley in the 1860s. Having lost two babies in Colorado, the young couple craved to start life over again. A wealthy landowner, Lucien Maxwell, was passing through Colorado and painted a rosy picture of northern New Mexico, all but inviting them to try their luck there. Roughly a year later, the couple appeared in Cimarron. Maxwell, a cross between a mountain man and a real estate broker, had not forgotten them. He offered them four acres of land for each wild horse they captured and broke—this is how Manley and Teresa acquired their first thousand acres! They chose their land with care, staying with friends until they found the right parcel, and later they slept in the old cabin of the infamous legend Kit Carson while building their own four rooms on the land. The valley they chose is framed by the Sangre de Cristos to the west and fed by Poñil Creek (presumably named after the native plant Apache plume or poñil, in the rose family) to the south. Protected from high winds by the mountains, and from colder freezes next to the river, the orchard sits in a prized temperate zone. The apple orchard went on to become a thriving business, but love was braided into it. It's telling that during Manley's later years, after Teresa died, tending fruit trees became his great joy. A telltale sign of a caretaker is a person who cares for the land and its trees primarily out of love and perhaps also to feed the dreamer within. I relate to this way of being, which is why I feel at home in orchards; caring for trees has a way of becoming a practice—we can't skip mulching or pruning and expect to evade the consequences—and, when the practice has ripened, the trees embrace us with such bountiful fruit.

The Chase family eventually acquired eleven thousand surrounding acres and went on to own the land for some 143 years. There were many good years for fruit, with 1879 being a highlight: "The apple harvest was the best ever. One variety, the Wolf River, was so large it took only forty-two to fill a bushel basket." Unremarkable as an eating apple, the Wolf River variety sings when it is baked or dried and is notable for its extraordinary size. In late September, when the Chase Orchard is intoxicatingly fragrant with apples, bears hover around the trees in broad daylight. "I can't get anywhere near the trees," Thelma said. "Because of the bears!" She is your superhero grandmother, somewhat wide and with an unmistakable sparkle, who serves on the board of the Chase Ranch Foundation. Some mornings, when she comes to the ranch, she also finds deer, elk, and wild turkeys roaming freely. She understands that this is their home too: "When humans aren't around, these animals take over." Not long back, a volunteer radioed the sole employee sorting papers in the old Chase house, asking her to stay indoors—because a mountain lion was roaming through the property.

History breathes among these ancient trees, and the pink adobe house hums with untold stories. "This is the tree under which Governor Wallace finished writing *Ben-Hur*," Thelma told me, referring to the novel scholars now consider "the most influential Christian book written in the nineteenth century." Lew Wallace was visiting Chase Ranch at the time, and his hosts Manley and Teresa deflected their neighbors' ardent desires to call on the governor, instead creating a quiet haven for him to finish writing his best-selling classic.

After walking back from the orchard, we stood near the Chase house, lined with ash trees that had turned yellow-gold. The pink adobe house, once a four-room dwelling, now sprawls with thirteen rooms. I gazed up at hillsides covered with oaks. Teresa and Manley had initially called their new home Oak Ranch "because of the trees on the hillsides that turned a soft reddish brown in the fall." Below us, in the distance, grew ripe-yellow cottonwoods. Abruptly, yipping sounds reverberated from the nearby Poñil Creek and saturated the air. Coyote pups! Everywhere the eye could roam, the olive-colored mountains gazed at us, and their beauty filled me to the brim.

"In the summer I have spotted the most beautiful summer tanagers I have ever set my eyes on, weaving along Poñil Creek," Thelma said. I lit up at the mention of the gorgeous rascally bird. It was by Poñil Creek that Teresa Chase, pregnant and with two babies in a buckboard, often gathered native wild plums, *"ciruelas cimarrones"* in Spanish, along with a friend and her baby—to make jam with. In the fall, some of the local families hunted together; at the time, thousands of antelope, deer, wild turkey, and elk ranged through the mountains and plains of Colfax County. Today, the numbers have dwindled—in proportion to the dwindled habitat—but the descendants of these animals still roam the area.

The sky was purpling and, in the growing darkness, we reluctantly got into the car and steered out of the exquisite land. I recalled the summer tanager, *Piranga rubra*, I had once seen next to a pond, green with algae, in Las Golondrinas; a foil to the croaking bullfrogs, the bird had lit up my heart. As we left Chase Ranch, I promised myself to return here come spring, to breathe in air scented with apple blossoms and look for the rosy-red fruit eater.

III

Lucien Maxwell, the landowner who enticed the Chases to Cimarron, had a keen interest in selling off parcels of the vast holdings he had inherited from his father-in-law. And he encouraged many homesteaders—as opposed to squatters—to settle lawfully in the area. Maxwell's sister-in-law, Petra, and her husband, Jesus Abreu, were also persuaded to buy land in nearby Rayado, a half-hour uphill drive from the town of Cimarron. The Rayado compound came to be known for the trading post that Maxwell and Kit Carson ran there, which was frequented by travelers on the Old Santa Fe Trail. Other structures likely included a blacksmith shop. Carson had a fraught relationship with the Indigenous peoples and was known for his friendship with compliant tribes and brutality against others, especially the Navajo. In 1850, a detachment of US Army Dragoons was stationed at this trading post to protect the local white settlers from what was seen as "Indian trouble." Conveniently, the white settlers forgot that this area had long been a cherished homeland of the Jicarilla Apache.

In 1857, Maxwell moved away from Rayado to found Cimarron, and the Abreus took over operations of the Rayado Rancho. After the Civil War, they began to plant what would become a commercial orchard. The next morning, a Sunday in early November, Dave Kenneke graciously met us at the entrance of the sun-washed Rayado orchard. Standing at the edge of this orchard, I observed a pear tree that had turned flaming red. The Abreus grew pears and apples, but their cattle brand was a pear, perhaps suggesting their preferred fruit. While the children climbed up a metal gate to gaze at a stunning wild horse, white as milk, Dave walked Michael and me past two ancient pear trees that had survived from the late 1800s.

We soon wandered into an old apple orchard, where three deer languidly munched windblown apples. Dave had had a hectic start to his morning—"Excuse me if my ears are red," he'd said—but his hearty face lit up when we stood among the historic Rayado trees. Mature fruit trees still thrive here from the days when the Abreu family ran the orchard, but young trees—Chase varieties grafted on to healthy rootstock—have since been added to the mix. Gordon Tooley and his assistants did some of the grafting. Gordon's father once worked for the Philmont Ranch in Cimarron, and Dave himself once worked for Gordon's father. I feel touched by how these fruit trees have braided lives across generations and go on connecting caretakers even today. Gordon's father is now at an assisted living facility in Albuquerque, but Gordon returns here occasionally, and he led a grafting workshop at Chase Ranch, which Thelma Colker herself admitted she greatly benefited from. "Chase's folly" has turned out to be anything but that.

The autumn air shimmered, as crisp as an apple's first bite. We sauntered for an hour in a flat section of the orchard, studying the fruit trees and occasionally reaching up to taste an apple. More apple trees were scattered along the river below us. A line of cottonwoods flared yellow, hinting at the curves of the nearby river. Dave recalled walking along the river with Gordon and finding an old apple tree felled by the wind. "But Gordon was able to salvage some whip," Dave said. He grafted the scion wood onto rootstock and, lo, a young apple tree was born. Later, Gordon did some research on the varieties he grafted here and was able to trace some back to Pennsylvania and Ohio—the saplings once came here on oxcarts in the 1800s along the Old Santa Fe Trail. Some of the older varieties that still bear fruit here are rare enough that Dave believes they might be the last of their kind.

"I do what I can to keep this orchard going . . . as a labor of love," he said.

While Dave is tied up with the day-to-day operations of the Philmont Ranch, he also oversees the restoration of riparian habitats along the river. "We are planting oaks and willows to grow alongside the cottonwoods," he said. The orchards are something he miraculously fits into his days. I asked what sparked his love of fruit trees.

"As a boy, I lived in northern Illinois. Once, with my Boy Scout group, I went into an old lady's orchard, pruned her pear trees, and when the pears came back stronger, my group sold them for a fundraiser." The experience spiked his interest in fruit trees, but his love of apple trees was seeded by his father, who periodically asked him to look out for the Northern Spy and Tolman Sweet varieties he had tasted in his Wisconsin childhood.

"Do you have a Northern Spy? A Tolman Sweet?" Dave's father would ask. Today, Dave grows both these varieties around his log cabin in nearby Miami, New Mexico. To grow a Northern Spy is a heroic commitment, for the tree takes ten years to begin fruiting.

Wide and exposed, the Rayado land is not sequestered between mountains like the Chase Ranch, but it is fertile land nevertheless. With pear trees being relatively easy to grow and apple trees thriving in this climate, the Abreu family operated their orchard here until the 1920s.

"Many of the apples weren't necessarily good to eat," Dave said. "But, like the Ben Davis variety, they were great for cider—soft or hard! Or they traveled well, to the East Coast, and on ships going across the ocean. And for these reasons they were considered to be keepers."

What is called the Native plum or American plum grew freely along creek or river bottoms in the 1800s—to Teresa Chase's delight, and mine. Today, wild plum trees grow along fence rows, benefiting from the lines along which acequias run; bears and raccoons carry their seeds over to propitious areas. Dave attested that these plums make wonderful jellies, especially when mixed with apple butter.

It was early November, but many of the apples at Rayado hadn't been picked. Having endured a recent frost, the large red apples on the young trees were no longer crisp; still, they gave off a heady fragrance. The midsize apples from the older trees, their skins softly tiger-striped red and yellow, had survived the frost nicely and were still crisp; their tart flavor, with hints of sweetness, was as refreshing as the sunny fall breeze. I recalled something Thelma had said: Her grandparents had told her that you don't pick apples until the first frost, for that is when they get sweeter. Thoreau, too, knew from experience that apples sweeten after a frost.

After the wild horse galloped away, the children joined us in the Rayado orchard. Michael found a smaller red apple on the young trees to have a savory flavor while the children demurred; they pronounced it plain tart. These November tastings brought the Rayado orchard to life, singing in the wild language of apples. I wondered if frontiersman Kit Carson, whose dark-red cabin stood unscathed next to the orchard, had once tasted the tiger-striped apples. The mature trees with wide canopies bore most of the apples, and they had achieved the size of a shade tree—the kind of fruit tree I love best. They imparted real beauty to the orchard that the young ones, fenced up to protect them against the deer, couldn't do yet.

When we returned home, I felt that searching for historic apples had brush-stroked my autumn with wildness; it lit up October and November in ways that made me feel buoyed by apples. The children and I had begun tasting them in August, maybe even in late July, when we were impatient for them to ripen. We might have been echoing Rilke's supplication: "Lord: it is time. The huge summer has gone by. . . .

Command the fruits to swell on tree and vine; grant them a few more warm transparent days, urge them on to fulfillment then." The August apples were green and tart but almost good as refreshment after a sun-baked hike. As September unfurled, the fruits slowly ripened, accruing flavor by the week until one weekend when the children declared that they were actually sweet. We began picking in earnest in mid-September and didn't stop for nearly two months. Hundreds of apples I picked vanished before my eyes, as happens when you share with children, friends, and neighbors. Kids who hadn't previously liked apples would stop by with requests for varieties they had fallen in love with: "Do you have that Russian Giant? How about the Deacon Jones?" Grownups from a carpenter to a car salesman were united in their admiration of the Jonathan, which has been called the essence of the American apple.

Now caretakers of our own apple trees, Michael and I gazed anxiously at their spindly trunks. In the afternoons, brutal windstorms snap branches and, in the mornings, a family of deer hover in the garden, eager to munch what leaves they can reach. We bought stainless steel fencing to protect the trees, and installed and staked the fences. While our trees couldn't be expected to bear a harvest anytime soon, those around me were noticing that they had an apple evangelist in their midst.

Chapter Ten

Industrial Fruit

NOVEMBER

Mia among apple trees. Photo by Priyanka Kumar.

I

Pika's baby teeth were falling out, and she found the dense red Winesaps we'd been foraging challenging to eat with her "wiggly tooth." She all but jumped for joy when John Trujillo pointed out a Golden Delicious tree in his family orchard in Chimayo, New Mexico.

"Oh, yeah, I love those!" Pika said. Sinking teeth into the lighter apple, she raced off to find Mia, who was playing with John's three outdoor cats.

John grinned. "I have a photograph of my mother in the 1950s, harvesting apples from this tree," he said, before stretching to pick more yellow fruit from a thickly dotted canopy.

The tree was laden with a few hundred crisply sweet apples and gave off a bracing scent. While the taste of apples is determined primarily by sugars and organic acids, their aroma "is a complex mixture of many volatile compounds" whose composition is species-specific and often distinctive in each variety. Scientists have identified more than three hundred volatile aroma compounds in the apple, but a team studying the biochemistry of apple aroma concluded that only a few of these contribute robustly to the fruit aroma: "They mostly include esters, alcohols, aldehydes, ketones and ethers."

The ground before me was littered with apples, and they hung from the tree like orbs of setting suns. I wasn't inhaling the heady fragrance of early-season apples or the sober aroma of mid-season apples. At John's orchard, the wind-blown apples were fermenting, scenting the air with cider. Which made sense, since the balance of volatile compounds in apples changes as they mature; at the beginning, aldehydes predominate, then alcohols start to spike, and finally the esters take over.

Since John hadn't harvested these apples for the farmers' market, I wondered if he'd saved this tree for his family's use. John's ancestors root him to this land: Almost seven decades back, his grandfather traded away a forested area north of here—because he didn't care for the snow—and in its place acquired this twelve-acre property in valley land. "In 1948, he planted the Stayman Winesap," John said, "one of the oldest crosses you can grow." A Kansas variety that dates from 1866, the Stayman Winesap was once prized as a dessert apple, and it exerted a pull on John's grandfather. He moved away from cattle ranching and began planting fruit trees instead. Now John tends to the trees.

Two stucco houses and a sprawling orchard stand on the property today. Few of us live on the same land where our grandparents, or even parents, once harvested fruit. The twentieth-century movement of people from the country to the city, and the relentless urbanization that ensued, sundered us from the land. Orchards were not exempt from this shift. In the post–World War II era, many were bought out by commercial interests. New, specialized orchards developed a hyperfocus on yields and sugary sweetness, attractive-looking fruit, and varieties that could endure long shipping days. In the industrial phase of our apple history, the obsession with sweet fruit and supermarket-ready skin spurred a modern reshaping of fruit, stoked by greed and sanctioned by the government. Roger Yepsen writes that a 1959 government guide to apple production actually "coached growers that 'sales may be increased 75 percent on the average by increasing the area of solid red color from 15 to 50 percent.'"

Americans got hooked on color. A ranger at the Los Luceros Historic Site in New Mexico told me that after seeing the big, red apples in *The Wizard of Oz* (1939), many Americans began to grow the Red Delicious apple. This is the dominant variety at Los Luceros too; now the orchard is aging, and those Red Delicious apples taste mushy. Regardless of whether people complain about the Red Delicious or tolerate it, the apple remains among the top varieties grown in America. For me, the apple's "lipstick red" color signals the loss of genetic diversity in commercial varieties and how during the industrial makeover, the soul of the apple and its marvelous complexity got lost.

John's Golden Delicious towered over us, teeming with shade and character. It is unusual to see such a generous canopy on a modern apple tree. Under the influence of the US Department of Agriculture, fruit trees also weren't allowed to grow to a mature height of five feet or taller, and natural tree shapes were discouraged. The exclusive focus on commercial fitness and high yields inevitably led to the varieties being grown to plummet—from many hundreds to tens. The aggressive mass production of fruits after World War II sounded a death gong for a plethora of fascinating apple varieties.

Commercial growers pumped the soil with unprecedented amounts of pesticides and fertilizers to stimulate ever higher yields. Their strategies did work in the short term, but decades later, we are realizing that conventional fruits and vegetables are grown at terrible costs to the earth and us. Americans now collectively use a billion pounds of pesticides and fifty billion pounds of fertilizers each year. Pumping artificial chemicals overstimulates the soil and ultimately deadens it.

In the fairy tale, Snow White almost chokes on a poisoned apple, and now our real-life apples are laced with toxic chemicals. American apples are sprayed routinely with diphenylamine, a chemical treatment that prevents the skin of apples in cold storage from developing brown or black patches known as storage scald. Because of their high pesticide residues, apples appear near the top of the Environmental Working Group's list of "Dirty Dozen" fruits and vegetables.

"Few Americans may realize it, but most conventionally grown apples are drenched in diphenylamine," the EWG cautions. Diphenylamine is banned in Europe, but American farmers continue to apply it liberally. European officials have concluded that diphenylamine can combine with nitrogen-containing compounds to form nitrosamines, which "cause cancer in laboratory animals, and some studies have found that people who eat foods with nitrosamines have elevated rates of stomach and esophageal cancers."

We pay a steep price for the apple's outer beauty.

John teared up when he spoke about how pesticide use—and working on a military base—sickened his father. "Eventually my father stopped spraying," he said. "The pesticides made him sick, physically." Struggling to speak, he grew silent. In the orchard, thick sunbeams washed over the trees. "I think he also got exposed to something . . . with the nuclear work he did . . . for the military. . . . He died of cancer." His face contorted with sorrow.

After that tragedy, John decided to pivot. "I take care of the land in the old ways," he said. Today, John calls himself a "natural grower," though you may not see him filling out the extensive paperwork needed to get a "Certified Organic" label. He speaks about how things were done in the 1850s—when the land was fertilized with manure, harking back to a time when there were few synthetic chemicals. A friend regularly delivers horse manure, which John spreads around his six hundred trees. He may not have an organic certification, but he no longer uses pesticides.

Near industrial orchards and farms where artificial fertilizers are used, phosphates from the fertilizers wash into rivers and then into the sea, leading to algae blooms and dead zones. Nitrate, mainly from polluted farm runoff, contaminates the water supplies of some 1,700 communities nationwide at levels the National Cancer Institute says could increase the risk of cancer. Among crops most doused with pesticides are cotton and apples. Are there better ways to dress and eat? Interestingly, scientists at the Graz University of Technology in Austria recently found that organic apples harbor a more diverse and balanced bacterial community—"which could make them healthier and tastier than conventional apples, as well as better for the environment." It is true that organic food is more expensive in the short run—and sadly remains cost-prohibitive for many families—but in the longer game, it nurtures our health and the soil. As a society, what do we owe the land, and ourselves? Can we find ways to pivot, as John and other organic farmers have done?

Over the last century, conventional orchards have become a monoculture of lookalike trees in unrelenting rows. The apple itself has gone from being a highly acidic fruit, prized for cider and dessert, to a cloyingly sweet supermarket staple. We have also lost our rich apple lore. Who knows anymore that the antique New Yorker, the Golden Russet (seedling of the English Russet), is the champagne of cider apples or that the relatively modern Winecrisp is excellent for pickling? That the Northern Spy needs to go into cold storage for a couple of months before its flavors develop fully, and the Wickson boosts pollen for cross-pollination? (In fact, 5 percent of what commercial orchards grow should be pollinator plants such as the Wickson). The loss of experiential knowledge—and an overreliance on pesticides—has repercussions for the rest of the ecosystem: Apple blossoms attract *Apis mellifera*, the honey bee, and other pollinators, including native or wild bees. Studies conducted since the 1970s have documented how important native bees are to apple orchards and how they face severe challenges from the use of pesticides. Scientists have observed that large apple orchards in China have lost their natural pollinators, most likely due to an excessive use of pesticides, and now rely on migrant workers to hand pollinate apple blossoms.

The arc of our apple story suggests that progress isn't necessarily a linear march. Early successes with apple growing were co-opted by agribusinesses, and the varieties Americans once celebrated are scarcely known today. Now our challenge is to circle back and think wild, to wander back to orchards that will nourish us instead of fattening corporations.

There are ways to grow apples without using chemical fertilizers and pesticides: Gordon Tooley recommends clean tools, disinfected with 99 percent alcohol before each use, to avoid bacterial and fungal infections. Pheromone traps can reduce coddling moths. Gail Haggard, who has been growing and selling native plants for half a century, has found that compost feeds trees more wholistically than any bottled chemicals—and healthy trees are better able to withstand environmental stressors, which are only intensifying. Using rootstock that matches climate and soil types, and avoiding apple varieties such as Gala that are susceptible to common diseases like fire blight, can also yield healthier trees. The US Department of Agriculture's Natural Resources Conservation Service has issued a new road map for sustained soil health that could be a golden maxim to move us away from toxic orchard practices: "Minimize disturbances, maximize biodiversity, maximize soil cover, and maximize living roots."

Unfortunately, independent growers like Gordon Tooley who care about genetic diversity and soil health have lost much of their business to box-store nurseries and barely eke out a living. To complicate the picture, commercial monoculture apple orchards are now being pummeled by disease. As climate change batters our orchards—John told me that most of his cherry trees have perished due to the warming climate—diversifying the genetic variety of fruit would lead to greater resilience to the multitude of threats the trees face. Global warming has cut into John's game, yet he's an outlier who proudly offers at least seven apple varieties at the farmers' market. While orchardists like

John Trujillo keep interesting varieties in circulation in markets across the country, apple diversification could happen on a more meaningful scale if we rediscover orchard culture and cherish the biodiverse forests that fruit trees once nestled against.

II

To get to John's orchard in Chimayo, we had to travel back in time. At least that is what it feels like when you are driving toward a dog sitting in the middle of a road, who is in no hurry to move as your car approaches. You know that you are in a tiny village—Cundiyo is the name of this one—with almost no traffic. Just a few paces ahead, another posse of unconcerned dogs laze on the road. Highway 503 weaves through sleepy Cundiyo and offers a rude awakening when it transforms into a mountainous road that curves and plunges steeply in the same instant. This narrow asphalt ribbon felt perilous as we drove in search of the orchard where the pears are infused with caramel notes and all but sing. Weaving past dizzying mountain roads, we got astounding views of what locals call the badlands—clay-colored hillocks, dotted with shrubby junipers, undulating across an infinite expanse.

After the adventurous drive, we arrived at the orchard later in the afternoon than expected, but John was waiting graciously. He'd asked us to look out for an orange-peach house, but there's more than one such house in this rural area. A man of big build, John wore a white T-shirt and jeans. A full head of graying hair was combed back into a ponytail, and a gray-white goatee framed his somber face. On occasion, his face lights up into a smile, revealing even, white teeth. He must be approaching seventy, but he stands straight and firm as though rooted to the earth.

Throughout their childhood and teen years, John and his brother helped their father plant many of the apple and pear trees that grow in the orchard today. In 1971, all these plantings were thwarted when the region suffered a big freeze. Temperatures approached minus 30 degrees and many of the old trees died. Today, only two of the oldest trees from the 1940s survive: a seventy-year-old Bosch pear and the Golden Delicious apple, both of which are very fruitful.

John climbed a ladder to pick four Bosch pears that had been overlooked during the last picking. Elegant fruit with Modigliani necks, their subtle caramel notes played over a buttery-sweet depth. Refreshed by the pears, we gazed at two rows of pear trees on land that inclines above John's grandfather's house. He is gradually renovating the house on his own, with an eye to living there. Across the dirt road from the Golden Delicious tree stood a battered car.

"I plan to repair it," John said.

A handful of old cars are scattered around the property, and one car he plans to "build up" is missing a front frame and is tenanted underneath by his three cats, who were wildly popular with my children.

After the 1971 freeze, the government gave farmers a deal to buy fruit trees from a nursery. "Everybody bought them," John said. "And my dad got his kids involved in planting trees." Fifty years later, temperatures are soaring in the opposite direction. When we had walked up to the pear orchard, John pointed to a large pile of "fruitwood"—skeletons of fruit trees that have perished, largely due to drought; the family will use their remains as firewood.

A decade back, John noticed that it was getting unmistakably hotter and drier in the orchard. Of his cherry trees, he said, "It is dry here, so the fruit doesn't get big . . . and my gold cherry trees died." He showed me one remarkably thick, dark stump. "The trees starving for water get weaker. And the bugs get in there—every bug in the world wants to eat them."

John had planted the cherry trees, including black cherries, in 1984—the year when he and his brother planted many new trees in the orchard. "There was a lot of water then, and a lot of rain." In the eighties, they also had Red Delicious apples in abundance. "When you have water rights, you have to use them," he said, "or else you tend to lose them."

Today he pumps water from the acequia into tanks that feed his drip irrigation system. A pear tree needs at least thirty gallons of water a week, and smaller trees need even more because the sun scorches them more easily. John told me that excessive heat paired with drought conditions is stressing fruit trees and making them terribly vulnerable to insects.

John has mournfully concluded that it is "too dry, too hot for cherries here." Then he added, "I'm not going to plant trees no more.... It's too hard." Now the temperatures can change by 40 degrees Fahrenheit in a single day, he said.

I felt aghast. John is losing hope. If veteran farmers like John Trujillo are deciding not to plant any more fruit trees because it's too difficult to keep them alive during global warming and drought, what hopes can the rest of us nurse? My butterfly self may have a vision to connect us to nature by way of fruit trees, but John has me wondering if that is still possible on land hit by climate change.

NASA scientists confirm the observations that John has made in his orchard: Since the pre-industrial period, human activities have caused the earth's global average temperature to increase by roughly 1.8 degrees Fahrenheit, and this number is rising by 0.36 degrees Fahrenheit each decade. Scientists know that human influence has warmed the atmosphere, ocean, and land. The Intergovernmental Panel on Climate Change keeps issuing stark warnings: "If temperatures keep rising many parts of the world could soon face limits in how much they can adapt to a changing environment. If nations don't act quickly to slash fossil fuel emissions and halt global warming, more and more people will suffer unavoidable loss or be forced to flee their homes, creating dislocation on a global scale."

Brazilian president Luiz Inácio Lula da Silva has said that in Brazil, the climate emergency is already a reality: "The Amazon region is going through an unprecedented drought. The level of the rivers is the lowest in 120 years. I could never imagine that this would happen in a place where we have the greatest reservoir of fresh water of the world."

United Nations reports confirm that nearly a quarter of humanity is now living under drought.

III

I would have liked to cross the main road and wander through John's six-acre orchard downhill from where we stood. For several minutes, however, we had been hearing gunshots coming from that direction at periodic intervals. Concerned that our children were floating around the lower edge of the pear orchard, I asked John about the gunshots.

"That must be my nephew, target shooting," he said.

A little later, the shooting had quieted long enough and it was on my lips to ask about going down to the orchard—but I heard the shots again, coming from where the apple trees grew.

"I've never heard them so close," John said. When his thirteen-year-old granddaughter, Candice, drove up in a motorized cart and asked if she could drive her three-year-old cousin to the apple orchard, John didn't let them go. The toddler looked impassively at us and blinked like a doll.

"You can only drive up around here," John told Candice. She did so with glee, grinning and showing off new braces and spewing gasoline fumes when she circled past us.

Walking us to a bare-bones structure in front of his grandfather's house, John said, "It used to be his greenhouse.... When I moved into this compound with my parents in 1967, Grandpa was growing a lot of chiles in the greenhouse and, all around the outside, *ristras* were strung up." In New Mexico, bunches of whole dried red chiles are arranged into festive *ristras* that can be hung up as decorations and used as ready spice.

While we strolled in the orchard, weaving through the pear and dead cherry trees and the old Golden Delicious, and when we conversed outside the grandfather's house, stuccoed peach (John's parents were peach growers), with Candice deliriously steering by, an old current of life coursed through the place.

After 1967, John would stay here for life, planting fruit trees, tending them, and selling the fruit. He doesn't care to process the fruit, though he juices it for his family. "I like to keep things simple by selling whole fruit," he said. Many orchardists I have spoken to have lives that are woven through with innumerable details, but at their heart Thoreau's mantra, "Simplify! Simplify! Simplify!" rings like a gong. Whether they know Thoreau or not, they are his brothers and sisters. A few days later, I ran into a woman from Chimayo who knows John and other orchardists in the area. I asked if she thought they sprayed pesticides on their fruit.

"No," she said. "They all get their water from the acequia . . . so they don't want to end up poisoning that water. All those old men, they have a . . ." she grasped for the right words.

"A feeling for the land?"

"Yes," she said. "Exactly!"

As a boy, John lived on a military base north of Tokyo, and each morning when he left his house to go to school, he saw Mount Fuji in the distance. The family later moved to the Kirtland Air Force Base (now a Superfund site) in Albuquerque and, on his way to school, John saw the Sandia Mountains, which felt like a mirror of his Japanese experience. In 1972, his father planted most of the trees (many of them Red Delicious) that grow in the orchard today under the gaze of the Truchas Mountains. A local co-op sold the apples for the family, but after some years the co-op collapsed. Then John's father met a Vermont man, who told him: "You should grow McIntosh and Pippins." McIntoshes are sprightly apples, and the term Pippin is used to describe several varieties, including the Newtown Pippin. John's father diversified the orchard and later got a Pippin variety from Gordon Tooley, whose farm is just a few miles north.

John told me with quiet satisfaction that he grows twenty-seven varieties of apples. On his six acres that hew the river, with the fence line flagged by a magnificent cottonwood, grow five hundred apple trees. I had a glowing view of the trees from where we stood, but because of the nephew's target practice we didn't go down there.

The afternoon was waning, and our children were still playing with John's cats. As we soaked in the view of the apple orchard from the grandfather's house, John told me something about his family history that stopped me in my tracks. He traces his paternal family line back to 1598, when the Spanish conquistador Don Juan de Oñate came to New Mexico, bringing with him soldiers and farmers—and possibly apple seeds and saplings. Growing food was on Oñate's mind: "On the 11th [of August, 1598] we began work on the irrigation ditch," he wrote, as Juan Arellano recounts in *Enduring Acequias*. It could be that the apples weren't initially successful, since the first fruit trees to become established in the area were apricot, peach, and plum. One of Oñate's men was a farmer by the name of Trujillo and would go on to become John's ancestor.

"My cousins have the paperwork to prove it," he said.

This is remarkable when you think about the fact that Oñate stopped in nearby Espanola in 1598, and a descendent of a farmer he brought with him is still farming within a fifteen-mile radius of the original site—more than 425 years later! While John acknowledged that Oñate was one of the more pitiless conquerors in this region, he also draws a distinction between soldiers and farmers and sees himself as a descendent of the farming lineage. John belongs to another significant farming lineage: those who choose to foster a relationship with trees instead of managing and colonizing them.

Chapter Eleven

The Archbishop's Garden

DECEMBER AND JANUARY

Paul Cezanne. The Basket of Apples, 1893. The Art Institute of Chicago.

I

A hundred-year-old Norway spruce was one of two guardian trees that stood in front of the Cathedral Basilica of St. Francis of Assisi in Santa Fe. As tall as the iconic cathedral, the sixty-foot spruce was known as the cathedral's signature tree. In 1922, a local banker would walk by the yard in front of the cathedral each day after work and decided that the yard looked barren—so he planted a tree there. Ninety-nine years later, a December day brought a windstorm with gusts up to seventy miles per hour—as estimated by the National Weather Service. The spruce, whose root system was likely dry from our megadrought, came undone during the storm and crashed to the ground. Soil moisture may be a significant reservoir of usable water on earth, but warming temperatures are drying soils to a degree that even established giants are falling. Fortunately, the falling spruce didn't damage the cathedral, and its boughs instead wreathed the statue of the man who had envisioned the cathedral, Archbishop Jean-Baptiste Lamy, while leaving him unharmed.

Johnny Appleseed may be our most famous apple evangelist, but there was another who worked just as prodigiously. French missionary Jean-Baptiste Lamy sailed for America as a young man and first began to preach in remote Danville, Ohio. A gaunt man with a square, determined jaw and a flash of light in his dark eyes, he eventually migrated west to Santa Fe in the mid-1800s, at the age of forty, and settled a short distance away from where I live today. Over the years, he planted a demonstration garden behind the Cathedral Basilica of St. Francis, which still stands like a pinkish-tan dream (Willa Cather immortalized Lamy's desire to build the cathedral in *Death Comes to the Archbishop*). Lamy's garden was nourished by a belief that a bounty of fruit can be grown in arid land. He brought in new orchard stocks at significant effort and expense—"ten or fifteen dollars" a tree and "ten dollars a pound" for freight. Lamy was a consummate tree-maker, and when his garden thrived and grew abundant, he would trim it for transplanting elsewhere. Using all methods at his disposal, he helped set out a thousand fruit trees in Santa Fe in one year alone. As Lamy's trees flourished, he initiated a second resurgence in fruit growing in the area. He would gesture to the long shady vista of his garden, gleaming with fruits and flowers, and say that the purpose of it all was "to demonstrate what could be done to bring the graces and comforts of the earth to a land largely barren, rocky, and dry."

Today, as a megadrought and bigger, faster wildfires turn this land even more barren, and unbearably hot and dry, I am struck by the prescience of Lamy's ecological vision. He must have had an abiding connection to the land to tirelessly plant so many fruit and shade trees. Some summers, when we get showers that revive our wilting plants, I wonder if the overall picture might somehow improve. I put this question to former state botanist Bob Sivinski.

"Are we out of the megadrought yet?"

"Well, we had one wet summer recently, but the winter preceding it was horribly dry," he said. "There was no snow. So, we sometimes have a wet season, but overall things are still dry."

At a time when the word ecology—from the Greek, learning about (logos) habitat or household (oikos)—wasn't yet coined, Lamy pursued his love of edible and shade trees and worked tirelessly alongside his gardener, Louise, to transform their stark landscape. When Lamy came to live in New Mexico, "there were almost no fruit trees, for the fruit culture of the Spanish and Mexican colonists had vanished," his biographer Paul Horgan writes. Lamy is said to have acquired much of his original garden stock from his native Auvergne in France, but he was soon adding to his garden cuttings from states as varied as California and Ohio. "From many a plains voyage Lamy had brought cuttings of fruit and shade trees and grape vines all the way in buckets of water, scarce as it was, to be planted on his arrival home." And he encouraged others to act similarly. Lamy liked to receive visitors in his garden, and there were many who came. While conversing, he would toss breadcrumbs in a pond, fed by a natural spring, where trout all but flew out of the water to catch the crumbs. "For a visitor, he could pick a peach of five and a half ounces, a pear of eleven, or an apple of sixteen." When he gifted a visitor one of his luscious peaches, "it was always with the request that the pit be kept and planted."

Lamy would have a profound influence on the local ecology. His biographer adds that while mid-century daguerreotypes of Santa Fe show almost no trees, in later photographs taken during the 1880s, the plaza and other streets have bountiful shade from maples, cottonwoods, locusts, and weeping and osier willows. Lamy's trees shed luster in the lives of many, from parishioners to friends whom he delighted by planting saplings in their yards. "One day his old friend Mrs. Flora Spiegelberg glanced out of her front window in Palace avenue and saw the archbishop planting with his own hands a pair of willow saplings at her front gate," his biographer writes. "When he was done with his spading, he blessed the young trees." It is a tender image. When was the last time that we gifted a tree—and planted it for the fortunate recipient?

Soon, Lamy was growing prize apples, which he distributed among students, visitors, and invalids. Like John Adams, he believed in the healing power of apples. When a parishioner, Major Sena's mother, was in bed with fever, Lamy threw on a shawl against the snowy day and slipped four apples from the garden harvest into his cassock: "They were rich in pectin and quinine, good for fever." At Major Sena's house, "he put them on the hearth to roast them, chatted with the invalid, and when the apples were ready, peeled them with his pen-knife and, slice by slice, fed them to the old lady. The family said it was 'a simple thing,' but they always remembered it."

My own education was seeded in similarly simple moments. Lamy's generosity recalled how my parents shared fruit, but other cues also informed my thinking. In Assam, when I wasn't wandering in a bamboo grove or looking for snake skins, I'd be swinging on a deep-green swing that my father had rigged from a backyard tree. Here, I experienced small insights that accrued over the years. At dusk, I might find my father standing out in the yard, next to the kitchen, slicing fresh ginger. I recall standing next to him and asking why.

"A slice of raw ginger keeps away tummy-aches," he said.

From the green swing, I once watched with horror as a man, who sometimes helped in the kitchen, severed the head of a chicken—but the chicken continued to walk and even run about, squawking madly, blood spurting, and quite headless. The surreal image stays with me as though it had happened yesterday. At the time, we ate meat maybe once a week. Over a decade after witnessing the headless chicken, I gave up eating meat.

In the summer, my father would come home for lunch and regularly add a whole green chili pepper to his salad. One afternoon, I asked if it was very spicy. He challenged me to try it—and eat it whole. I hesitated. The peppers in this part of the world are famously hot, with the ghost pepper (so named for vaporizing the eater?) shattering the Scoville scale. But then I saw that my father was smiling, and I trusted that this would lead somewhere. I began to eat the raw pepper, tiny bite by painful bite, eyes streaming with heat, brain partially vaporized. It took some time before I understood why he had asked me to do this. My taste deepened, moved away from sugar, and I came to relish experiencing food close to its elemental form: fruit straight from a tree, mint leaves picked and pounded into a delicious chutney, raw ginger with a pinch of rock salt as an ally, and even the snakebite of a scorching pepper. Bite by bite, my consciousness grew alert to the pleasurable, medicinal, and eye-watering ways in which nature nourishes us.

The land of Lamy's birth is still known for its ancient and rare apple and pear trees. France is a major apple grower, with Normandy alone being home to over eight hundred varieties. The French laud their pomme de Limousin, a green apple "with a reddish hue" as having "a unique combination of sugar and acid, making it the perfect accompaniment for sweet and savory dishes." Lamy spent his childhood in Auvergne, where the fruit-growing culture would have fostered his love of fruit trees. Today, Auvergne, along with Provence, is the third-largest fruit-growing region of France, with seventy-seven thousand acres dedicated to fruit cultivation. Provence memorably gave us Paul Cézanne, who worked feverishly to create living apples on the canvas, apples that the poet Rilke would laud as an illustration of how art can produce "things" that are eternal. In letters to his wife, Rilke remarked on the "humbleness" of Cézanne's objects: "the apples are all cooking apples"; out of these everyday materials, Cézanne would make his "saints," Rilke wrote, and force them "to be beautiful, to stand for the whole world and all joy and all glory."

Fruit trees remained Lamy's lifelong passion. In his later years, he purchased a property north of town in Tesuque with an eye to the fruit-growing potential of the land. Some of his Tesuque apple trees continue to bear fruit today, though the property has long since been converted to a luxury resort, Bishop's Lodge, with rooms starting at $700 a night. I hunted down a hotel manager and asked if she could tell me anything about Bishop Lamy's trees. She didn't know what I was talking about but responded with warmth. "Talk to our gardener, Bobby. He knows everything."

I nodded, feeling a glimmer of hope, but couldn't get ahold of Bobby.

In December, I walked along the Santa Fe River until I reached downtown, and I wondered what our cities might look like if the culture of loving trees was restored. What remains of Archbishop Lamy's once-beautiful garden in downtown Santa Fe are not even a handful of dying trees. I can't imagine a heritage more worthy of restoring. An appeal was made to church officials more than a decade back, apparently to little avail.

Just when we need trees more than ever before, many species are growing endangered: The International Union for Conservation of Nature Red List of Threatened Species states that out of some 47,282 tree species assessed, more than a third—16,000—are threatened with extinction. Planting trees en masse may not be feasible in a desert, but it is possible to create microsystems of shade and bird habitat in our yards and streets by planting native trees. Hermann Hesse saw trees as penetrating preachers: "I revere them when they live in tribes and families, in forests and groves. And even more I revere them when they stand alone."

II

As I walked through American apple history, strands from my own apple past began to vibrate. On New Year's Day, I called an aunt in India to inquire about her health, and she recalled the time when she had visited us in our remote mountainous home in Himachal Pradesh—where I'd spent my early childhood. "I came because your father got a promotion," she said. "And they wanted to have a little party. I stayed a few days with you, in the mountains. Every morning, I watched you and your brother get on the school bus—with an apple in your hand. You always had an apple in hand!" It had been some time since I'd spoken to my aunt, and she knew nothing about my deepening interest in apples. I felt my arms prickle when she, on her own, brought up my early obsession. After I put down the phone, memories crashed over me. In those years, I had lived in the land of apples, and fruit trees were woven through my everyday life. It was as though I could see film footage of myself as a child weaving through apple trees.

After Lamy's death in 1888, his five-acre garden was tragically neglected. An 1887 photograph of the archbishop's garden shows a wide carp pond surrounded by thick willows; a "duck pond" remained on site until 1915, but then the Santa Fe River was dammed, drying up the spring that fed the pond. By the 1960s, the orchard garden had turned into a "trash heap," and in 1984, two sons of local politicians bought the land for a development project, razed the "debris" and, when money fizzled out, simply "closed up" the land with asphalt.

A contemporary photograph shows the site to be all asphalt and SUVs. It pains me to walk through the parking lot that the archdiocese has since put in where Lamy's glorious garden once stood; the ravaged fruit trees are almost all gone, and asphalt smothers the very soil Lamy once nourished. I recalled reading that Lamy's strawberries were so spectacular that he sold them for a dollar a box and donated all the proceeds. The summer after Lamy's death, his garden yielded "fifteen hundred quarts of strawberries, forty gallons of cherries, one thousand [gallons] of currants, and two hundred [gallons] of raspberries; while five thousand shrubs, vines, and young trees which were ready for transplanting from the garden were auctioned for charity in the plaza of Santa Fe."

What remains at the edges of the asphalt I stood on is a dying almond tree and a couple of neglected pear and mulberry trees. I pointed out these trees to my children, and they looked on blankly as though there was nothing left to see. They had wanted to join me on this fruit tree hunt, and I was sorry to disappoint them. In the end, we crossed the road and walked along the Santa Fe River, where we found a volunteer apple tree growing. I craned up to find an apple or two for the kids and wondered if this tree had some connection after all to Lamy's original garden. The archdiocese long ago sold off adjacent lots to the Drury Hotel Group and other commercial developers. Lamy's brilliant ecological vision seems to have found no purchase with church officials; the solace and inspiration that many of us might have drawn from the archbishop's dazzling garden is irretrievably squandered.

In the end, shortsighted greed and a ravenous lust for more development are what swallowed Lamy's garden—and this land lust has far from ebbed.

III

There is no better way to strengthen a community of apple trees than to become a tree maker—then you can graft a tree for your own garden or transplant it for a friend. One snowy winter morning, I left the house early to drive to a grafting workshop with Gordon Tooley. On the drive over, seeing the primeval landscape incandescent with snow, I got lost in the silence and didn't want it to end. It feels like a gift when I can still experience the landscape in its immensity. The experience ended when the Honda got stuck in a massive block of icy snow at the entrance to Tooley's orchard. Fortunately, an old man whom I'd never seen before ran over to help and, a little later, I found myself standing in Gordon's shed, apprenticing in the ancient practice of propagating fruit trees. The daylight hours were spent standing: cutting whips; aligning them against rootstock while keeping as much cambium contact as possible; gluing them to rootstock; taping grafts in place with a rubber strip; painting the exposed top of the cultivar with a grafting sealer; labeling the cultivars; and storing them in a wet bucket. The process felt workmanlike and required real precision. But a graft that takes makes magic. For you are creating new apple trees! I was eager to grow the Canadian Strawberry, a hundred-year-old apple variety from Maine with magical colors: "Skin color is buttery yellow with some green background, halfway covered with stripes and spots of vibrant red-orange."

Gordon Tooley wants trees that can stand on their own and don't need staking, as opposed to dwarfs, which need staking their whole lives. He uses an 890 Rootstock, which may be precocious but is resistant to fire blight, a significant problem in the area. "Grafting can be deployed to renew a shy bearer, or a tree which hasn't borne in years," he says. There are hundreds, even thousands, of ways to graft, and Tooley taught me three useful ones: the splice (a paintbrush cut made in three bold knife strokes), the cleft (better suited for bigger trees), and the bridge ("powerful medicine")—which features a "church window" cut and, if you're not careful, can lead to a very bloody finger.

When I brought my grafts back home to overwinter them in a bucket of moist woodchips, it felt like the beginning of an adventure. The first grafts survived until the next spring only to give up the ghost after they were planted. I was terribly disappointed, but there was one consolation: I had joined a lineage of tree makers stretching back to 2000 BCE.

Chapter Twelve

Community Making

FEBRUARY

A Lady's Blush, A Spitzenburg & A Crab Apple, ca. 1850–1900. Waxman Collection of Food and Culinary Trade Cards, #8573. Division of Rare and Manuscript Collections, Cornell University Library.

I

To strengthen a community can seem a nebulous task, but nothing is more essential in the individualistic world we live in. Lamy's legacy had not entirely faded, for it lit a candle in my heart, and I felt encouraged to take a research trip to Arizona and into the heart of a garden that is believed to represent the longest continuously inhabited area in the US—archeologists have documented some 4,100 years of continuous civilization. At the birthplace of Tucson, I studied an orchard where grafting has meaningfully brought together history and the present; my recent deep dive into grafting gave me a clearer understanding of what an ethnobotanist has accomplished at the Mission Garden. Exploring the orchard, at the base of Sentinel Peak, was like walking through a Southwestern fruitery. Jesús García's grafting experiments have produced 167 heirloom varieties, which coincidentally echoes the number of varieties in Jefferson's fruitery, and 36 grapevines. Led to historic sites in Southern Arizona by wanderings or hearsay, García collected most of the stock for these trees and vines—initially grown by Jesuits and Franciscans, from the seventeenth century onward—at old or abandoned homes, the University of Arizona campus, and even at the Quitobaquito Springs in the Organ Pipe Cactus National Monument. The trees were then propagated with assistance from the Kino Heritage Fruit Trees Project (named after the Jesuit, Father Kino, who introduced fruit trees—peach, quince, pear, fig, pomegranate, and apple—to the area in a transformative way in the late 1600s). The new plantings strengthened a living community of historic trees at the Mission Gardens and at the orchard of the Tumacácori National Historical Park, forty-five miles south of Tucson. I got to Tumacácori a week too early and the apple trees weren't in bloom yet. But a *Pyrocephalus rubinus*, vermillion flycatcher—a male in pomegranate-red and black plumage—flitted among the fruit trees. The pomegranate trees in this orchard also have an ancient lineage, and they bore abundantly last year—the fruit was distributed in the community. The shapely flying pomegranate—with a pirate's eye patch—followed me around as

I wandered among fruit trees readying themselves for spring's arias; this flycatcher loves being close to water, and soon I was following it to the nearby Santa Cruz River.

When I arrived at the Mission Garden, García was busy and somewhat tense as he prepared to lead a school group. I walked through the fruitery on my own and, later, we all met up in a pagoda, where I watched the schoolchildren try the cacti, rare oranges, and other awe-inspiring fruits García offered them, and me. Seeing the schoolchildren glowing after the unusual tastings, I wished that Pika and Mia were with me.

García's zeal recalled how my father poured his free time into taking care of our fruit trees in Assam—as a child, I watched meditatively as he and a helper grafted or pruned our peach, plum, and jackfruit trees. I have often wondered if my father inherited the ethic of nurturing trees from the failed farms of his childhood. During my childhood in Assam, harvest time was a grand occasion when I watched my father, and the men he'd corralled, leaning ladders against impossibly tall trees and climbing down with jackfruits as large and spiky as a dragon's head. The air would be charged with expectation. The tropical climate we lived in fostered an abundance of fruit, and my father would pack dozens of banana bunches in our cellar until all the walls were lined with bananas. Since the cellar was where I retreated to think or read, from an early age my reading was infused with the fragrance of fruit. My father took care that none of the remaining fruit was wasted. Handwoven jute baskets laden with fruit were distributed throughout the community, and this is the time when he would smile more than usual and his eyes would gleam with contentment. In our house, fruit that couldn't be stored for the winter was boiled or marinated into jams and pickles. My mother was a gifted educator who went on to become a school principal; it wasn't until after she had passed away that I learned from old neighbors in Assam that her fruit jams were legendary. Bottles of jam and jelly were shared all around, invitations dispatched for afternoon teas, and the expectation in the air ripened into something richer than the delectable tastings. I was seven when I watched intently, and with wonder, as the work of harvesting and processing fruit transformed into a community-making endeavor.

This work continues at the Mission Garden through caretakers like García. Here is a way of life that honors community. After the Tucson schoolchildren left, however, García grew somber. He told me that the rich contributions the Spanish have made to fruit growing in the US aren't acknowledged, much less celebrated. I recalled that over four hundred years ago, when the Spanish began moving northward on the Royal Road of the Interior, they brought apple seeds with them and planted them actively. While many apple trees from that era have waned due to senescence—most apple trees rarely live more than 150 years—some mature trees persist, albeit in grafted forms. It is now established that the first apple trees in the Southwest were set in the ground by a Spanish rancher north of the Salinas Valley as early as the 1630s (soon after, settlers from England, including William Blackstone and others, started planting apple seeds in Massachusetts), but for some reason this history isn't anywhere near as well known as that of the settlers.

"They [the rangers] don't talk about the pre-1850s history of fruit," García lamented. "Why is that? The Spanish made these amazing canoas to connect water sources and irrigate fruit trees, but no one knows about it!"

It struck me that there was a key question at the heart of García's lament: Whose history is amplified and whose is all but forgotten? Widening our apple history could help inform how we reshape our ecological future.

Back in New Mexico, I walked with fresh appreciation along life-giving acequias dating from the early 1700s. Oñate's men and the Puebloans he had no doubt coerced had begun digging these irrigation ditches as far back as 1598 and planted apple seeds nearby. Going further back, Arabs introduced apples to Spain in the tenth century, and they called the fruit tuffah. I felt surer that knowing the rich culture of apple growing is a pathway to caring about apple biodiversity, and biodiversity at large. I also wondered why, after the Spanish made so many lasting contributions, did the fruit-growing culture in the Southwest wane? Why did it take the passion of Jean-Baptiste Lamy from Auvergne, France, to revive our orchards?

II

On my daily walks near our home, I sometimes stop to speak with a neighbor who has a fine herb garden: an elegant woman who is originally from Guatemala, Consuelo, along with her husband, Dwight, bought a house in this neighborhood in the early seventies. As I'm leaving their garden, Consuelo might hand me a cutting or bundle of herbs. A couple of years ago, she pressed into my hands a cutting of Mexican oregano, which thrived, despite periodic neglect, in my terrace—and is now routinely sprinkled into our wild salads. Pondering these verdant gifts, I began to make fragrant rosemary bundles for Consuelo, who hasn't had success with this herb; whereas the rosemary I put into the ground a dozen years back is now a massive mother bush who is sprouting young ones. In April, our rosemary flowers are softly blue and native bees hum and hover over them all morning long. Stimulated by the gift of rosemary, Consuelo gave me a tour of her constellations of herbs and shared lamb's-quarter, basil, and varieties of oregano and parsley—all of which I experimented with in salads and stews.

The trove of herbs I was propagating from cuttings grew exponentially when nursery owner Gail Haggard gave me an open hand in her herb garden. Sunny and astringent, Gail has a deep ecological consciousness and, over the years, I have relished her presence. I once invited her to Mia's birthday and fed her homemade chickpea soup; another year, Gail brought her son and grandson (who is Mia's age) who were visiting from the East Coast to our home. The reward of Gail's friendship has outstripped my hospitality. It was nearing the end of the season, and the herb garden in Gail's native plant nursery was slated to be pruned. One fall morning, she handed me a pair of scissors and brown bags and implored me to take what I wanted. "All the plants will need to go into the ground before winter," she said.

So, I snipped away and filled up paper bags until they overflowed with all shades of green and were haloed by purple oregano flowers. I wonder why I feel exultant in such moments. A few months later, I came upon a passage in E. O. Wilson's *The Diversity of Life*, which revealed that the lower slopes of the eastern Himalayas—the area in

Assam where I had spent the formative years of my childhood—are considered by scientists to be one of the eighteen most valuable biodiversity hot spots on our planet. Only then did I fully appreciate that as a child I had lived in remote areas teeming with ten thousand plant species, some 40 percent of them endemic.

When I first moved to Santa Fe, it still felt like a small town. Things have changed since, but I wasn't too surprised to hear that Gail is also a dear friend of Gordon Tooley. "I recently spoke to her about my financial struggles running my business," he told me. I felt disturbed when Gordon added that Gail is "killing herself" running her two nurseries with their many employees.

Gail's and Gordon's troubles stem from systemic issues. At a 2009 conference in Madison, Wisconsin, a group of apple conservationists noted with concern that "roughly nine out of ten apple varieties historically grown in the US are at risk of falling out of cultivation and falling off our tables." Addressing the reasons for this decline, they agreed that one driver is "the demise of independently owned nurseries, which have had their business usurped by the garden-and-lawn departments ('pseudo-nurseries') of big-box stores." Just as many book lovers have chosen to support independent bookstores, plant lovers could engage in community making and bring independent nurseries back from "life support."

"What are we going to do?" Gordon asked Gail.

"We'll combine forces and crawl to the finish line."

This is classic Gail gallows humor. In her eighties, she can be seen skippering about whenever I visit her nursery. If there was ever a person who is having a fabulous time while "killing herself," it must be Gail. She may have a somber strain when she reflects on how we are poisoning the earth, but she also exudes cheer that bubbles up from deep within.

Armed with Gail's herbs, I was soon sprinkling fresh tarragon and three types of alliums into sautéing mushrooms, and I discovered two more varieties of oregano, though one is decorative. Until late autumn, I gave away portions of Gail's and my own herbs, and pots of strawberries I had propagated, to many friends and acquaintances. In so doing, I experienced how sharing food can nourish a sense of community.

The late Neem Karoli Baba, the spiritual force behind the Hanuman Temple in Taos, often said, "Food is God. Feeding the hungry is actually worship. God comes before the hungry as food. First bhojan (food) then bhajan (prayer)." I couldn't give away baskets heaped with tropical fruit to the community as my father had done in Assam. Not least since I live with severe drought whereas tropical Assam is one of India's rainiest states, with an annual precipitation of some seventy-six inches. But it gave me joy to hand out paper bags of apples and assorted herbs to those who walked up to our pórtico. Whether I was seeing friends and neighbors outdoors or indoors, now these visits, flavored with herbs and apples, grew more rasa filled—and the community they fostered felt real.

Chapter Thirteen

Mapping the Orchard

MARCH

Author labels apple trees at the historic Rayado Orchard.

I

The weather can be mercurial in the second week of March, with the sun shining one day and snowfall expected the next—but pruning season has begun. Two and a half years after my first visit to the Rayado Orchard, I returned, this time with Gordon Tooley and some other apple growers to sharpen our pruning and caretaking skills. Having pruned with Gordon at the Real Orchard, I had internalized his guidelines, and we soon set to work. I found myself with a nine-year-old apple tree and shook it to check if it was anchored well and if its branches rattled very much against each other. The tree was anchored, which meant that pocket gophers hadn't gnawed up the roots, but it sure was a rattler. I would quiet this rattling by pruning back the branches that were growing inward—toward the center of the tree instead of moving out laterally. Essentially, I entered into conversation with the tree, directing its branches outward to minimize competition for air and sunshine. Going by old pruning scars on the limbs, I could all but see the cuts that Ernesto, the retired groundskeeper, had made in years past.

"We're mapping Ernesto by old pruning scars," Gordon said. He'd told me at breakfast that Ernesto, who retired last year, pruned too heavily. But then we all have our own ways.

To begin with, I cut off the hardware cloth that was wrapped around the base to prevent mice from girdling the young tree. Then I cleaned up the western wheatgrass around the trunk with a trowel so that the base—the trunk flare—would stay dry. Gordon had told us that mulching right up to the trunk flare encourages wetness there, which can lead to decay and, eventually, lack of vigor—so the floor of fruit trees must be cleaned once a year. The rich dark soil underneath the wheatgrass was enviable. But this wasn't time to pause or gaze. Next, I snipped the water sprouts or suckers as close to the ground as possible—to discourage them from sprouting again—tearing them off with a hand snap, cutting with loppers, and getting a pesky one with a saw. It would be a worthy undertaking to prune annually the detritus of our lives with such verve. Once the circumference around the tree base was clean, I cut the suckers into smaller bits with a hand pruner and used the twigs to create what is called an "eyebrow" near the base to detain rainwater and sheet flow and serve as habitat for spiders and other beneficial insects. I finished my work around the tree base by covering the eyebrow with grass.

A *Sialia mexicana*, western bluebird, flew past me and perched on an adjacent apple tree. Sporting a cobalt back and lush chestnut breast, the bird watched me obliquely. A boon to the orchard, bluebirds fly-catch insects who are considered to be orchard pests. So, bluebird boxes are an intelligent addition to any orchard. Gordon adds that it's even better if bats hunt moths at night—bats are known for eating 40 percent of their weight in insects every night—and if owls keep a lookout for rodents who can destroy young trees. I thanked the bluebird for making me look up. In the glinting morning light, I relished the view of the mature oak thicket beyond the orchard, where the river races and snowcapped mountains crown the landscape.

Refreshed, I disinfected my pruners and prepared to cut the busier branches and the ones growing inward. Opening up a tree for light and circulation leads to an even ripening of the fruit. "It's good scaffold building while the tree is young," Gordon said. I cut close to the bud to stimulate branch growth, executing clean cuts that didn't leave a lot of surface area exposed.

Gordon had encouraged us not to hurry: "This is permanent agriculture, it will take the rest of your life, you can't be finished by noon." Instead, I pruned intuitively, "without pain or guilt." That being said, in the dry Southwest, it's best to prune no more than 20 percent of the tree.

I spent the morning reading apple trees as though they were the pages of a book, with the orchard being the storybook. I had begun to read fruit trees intuitively before I began to read books. As a child, I would quiet internally to sense the spirit of a tree, the feel of a grove, and spend hours in the companionship of trees. At the Rayado, I felt kinship with the trees and especially bonded with the old Duchess, who had been growing here either since the 1870s, or maybe a decade later, and was still sending up lively shoots. My fingers traced where her maternal trunk had been scratched by climbing bears and probed her ample hollows, where generations of birds and small mammals had raised their young. A red-shafted flicker had poked little holes along the Duchess's circumference—to suck sugar from her inner bark with a long tongue. After lunch, when I would begin mapping the orchard, I kept my roll of labels in the crook of the Duchess's maternal arms. Her touch might bless the younger trees in the orchard.

Some years back, Gordon had drawn a map of the trees he and a couple of helpers had grafted here. In preparation for this trip to the Rayado, however, when he'd searched for the original map, he couldn't find it. After considerable anxiety and an intensified search at home, Gordon at last located his nine-year-old map—on three pieces of scrap paper torn from an old notebook, and taped up in places. On the March afternoon, I studied the fragile map while the paper threatened to go to pieces in the wind. Gordon and I stared at the markings—crosses and abbreviated names—for some time, deciphering tree names and identifying landmarks: an old walnut tree, a fire hydrant, and the grand Duchess herself. After decoding Gordon's map, I began to create labels for the trees that had been grafted in 2017 using historic scions.

"Documentation is critical," Gordon said, while owning the irony that he had almost lost these scraps of paper.

Over the course of two days, I would label the trees and gather origin stories, so that this arboreal history will become accessible to generations who might one day wander through the Rayado Orchard. As I labeled the trees, I felt that I might have been giving nicknames to friends. At the end of the first day, I showed Dave the location of a beautiful tree named after him, Kenneke #5—the scion was taken nine years back from his yard—its bark glinting, planted next to the old Duchess herself. He smiled delightedly. I hadn't doubted that the day's work was worthwhile, but seeing the light in Dave's eyes, and recalling how he'd once introduced me to this orchard, all my tiredness dissolved. When I'd first come to Rayado, I was struck by the history, juicy and bitter, that had transpired here and couldn't have imagined that I would one day find myself mapping this orchard and that these trees would wear tags with my handwriting; it felt like having an undeservedly intimate connection with the trees.

Gordon had planted a row of apricot trees near the mature black walnuts—the walnuts secrete acids from their roots, and young apple trees don't like to live close to them. Other older trees, the mother trees, in the orchard include the McIntosh, Ben Davis, and the Duchess of Oldenburg. And the glorious Abreu pears from which the younger ones I labeled were grafted. I imagined that it would take another season of work to go beyond surface names and decode the apple varieties. Some were evident—Baldwin, Holt, Ben Davis—but other names simply described where the scions had been taken. One tree was named Amazing Regrowth, the scion having been taken from an old tree with one shoot coming out, maybe its last one. I felt like I was in a library of rare books, some of which had lost their titles. I was labeling some spines with actual titles and others with descriptive ones, so that a future scholar might have a way of orienting herself. A tree named Roberta Moore was grafted from a fifty-foot tree in Embudo; the owners were going to "kill the tree" for new construction but asked Gordon to save it by making some grafts, which he did, and planted some here. The scion for the Bear Tree was taken from an old tree at the end of a culvert in a nearby forest where a bear would regularly cross the highway. A rich pile of cobblers circled the tree. Sorting through these micro-stories, the Tuesday at the Rayado passed swiftly. I planned to be back early on Wednesday.

"In a few years, you can come pick here with the girls," Gordon said.

"I'll be here before that," I said, walking across grass bleached tawny. As it turned out, I would return to the Rayado a couple of weeks later, for the children's spring break. During the pruning and labeling session, I was staying by myself at a local ranch, and the children were at school. In the evening, when I spoke over the phone to Mia, she said, "Ms. Cherry told me to tell you that she's pruning her apple trees now, and applying that white stuff on them."

"Oh, yes, Kaolin clay paint. Thank you for letting me know. Did you mention that I'm at the orchards in Cimarron?"

"Yes, I told her."

I thought that the next time the girls and I harvest apples at the Rayado, it would be nice to bring some back for Ms. Cherry, Mia's teacher who shared her superb Black Oxfords with us last October. The Black Oxford skin was deep dark red, like the Double Red variety and the Arkansas Black. The flesh was whitish green, with a slight nutty taste, reminiscent of pistachio. Michael felt the taste suggested a squash or pumpkin. To thank her for the Black Oxfords, we sent Mia back with two varieties—Maiden's Blush and Rome Beauty—for Ms. Cherry.

The brief conversation with Mia moved me, but it wasn't until the next morning when I woke up to the cries and sight of wild turkeys that I figured out why. My daughter and I are talking apple trees! She is stepping into the world of fruit trees and growing familiar with its lexicon. Which is fitting since she might one day find herself taking care of the fruit trees that Michael and I have planted.

It fills me with rasa to see the children develop their own relationship with apples. In the way that I've been mapping the Rayado Orchard, Mia has also drawn a map of apple trees in her consciousness over our years of foraging together. She cross-references this apple map when she asks a question about ripening time or expresses a preference for a specific variety, whether it is the Deacon Jones or the Baldwin. In our family, Pika is famous as the discoverer of the Deacon Jones. One fall, she picked the largest apple in sight in Gordon's orchard. Naturally. The crisp and impossibly juicy apple became a hit with our girls and their friends. Our poor cat, Coral, found her name temporarily changed to Deacon Jones.

The Baldwin variety also won the girls over with its deep red skin and white flesh. Baldwins sweeten beautifully on our counter. Biting into one in November puts me into a bright wintery mood. Now the girls' friends, too, talk apples with me: "Remember that time when you cut us the Russian Giant?" with a twinkling smile. This passing on of apple love to a cohort of children, between the ages of eight and twelve, feels more important than words can express. "I never liked apples," Mia's friend told me. "And both my parents knew that. But things changed after I started eating all the different ones at your place." I found her newfound interest in apple varieties to be more gratifying than she realized. A few weeks later, Mia's friend asked me when we might go apple picking again. We recalled happily how, last fall, I had taken this girl and her younger sister to Los Luceros to pick apples. In my mind, images flashed of four girls racing through a sunny orchard, using a bamboo picker to snag the apples that were high up, and eyeing each other with horror when a crabby old lady tried to swipe the picker from under their noses and told Michael to "go away" for letting the children use the picker first.

At Wednesday's breakfast, Gordon and his two part-time helpers were huddled over iPhones, comparing the percentage chance of rain and snow over the rest of the week. "I'm on NOAA, and it says 40 percent today," Gordon said. A snowstorm was supposed to hit Cimarron that night, and it was already snowing heavily in Truchas, as Gordon had learned from his wife, Margaret. My own family has only one car, and I had planned to carpool back with Gordon to Truchas—we were to leave directly from Chase at around 5:00 p.m. If all went well, we would reach his Truchas farm before 8:00 p.m.—and Michael and the girls would pick me up and we would drive forty-five minutes south to Santa Fe.

The day still lay before us, and we would be out in orchards the whole time. It was a delicious thought. The air outside was bracing, and I layered up. At the Rayado, beyond the thick oak grove with crisp and russet leaves, loomed a few more bogglingly old apple trees. They might not live much longer. The morning grew colder, affirming that a snowstorm was on its way. I went on labeling the trees, and I sensed that I had fallen in love with them. At one point, my hands grew stiff with cold, and I ducked into the passenger seat of a car to write out some of the labels. As soon as I thawed, I was moving about in the orchard again, reorienting myself to one row or the other, cross-checking the trees against the deteriorating map. When I was labeling the young trees, the mature trees seemed to hover over me like elders. They watched my work, and old branches seemed to nod approvingly.

In an orchard of some 350 trees, and almost no tags, it was heady work to gaze at Gordon's scrap-paper maps and work row by row to tag each tree that I could identify and name. I was only tagging the young trees planted since 2012 but, an entire day later, I was barely halfway done. We had to leave by noon; however, I held a roll of labels that I hadn't yet attached to certain trees. Since I hoped to return to the area in a couple of weeks to look for migrating birds, I decided to complete the labeling then.

Before lunch, we hurried over to a private ranch across the road to search for a very specific tree. Several years back, Gordon and Dave had taken a trip to the Riviera Mesa to get a scion off an old apple tree. The trip through wilderness to the mesa all but broke Dave's jeep. Once they got there, Gordon found the scion he wanted. But before they could leave, Dave experienced an acute attack of back pain. "I thought I'd have to leave him there, and cover him up with some rocks," Gordon joked. "It's as good a place as you can find to die."

Dave, who has since had back surgery, chuckled grimly as Gordon told me the story.

Not long back, a wildfire ravaged the wilderness area, and Gordon suspects that the original Riviera apple tree may have burned down. So, it was important to check whether the Riviera scion he'd grafted onto rootstock for the private UU Bar Ranch had survived. Many of the fruit trees Gordon and his helper, Kelsey, had grafted here in 2017 were still alive, though badly in need of pruning. The grafted Riviera tree, unfortunately, had not survived.

The ownership of the ranch has changed. From a private owner who'd wanted an heirloom orchard, the ranch had gone to corporate ownership, and the way the grounds crew cares for the trees seems to have changed as well. A Goat-Camp pear, a variety from the turn of the last century, and a Dawson crabapple—the Russian variety called Dolgo, which means "long"—have survived. But the Riviera apple is certainly dead, and its backup in the wilderness has almost certainly burned down.

II

By the time we got to the Chase Orchard, the sun was playing hide-and-seek with us.

"This is ground zero for fire blight," Gordon said, examining some young apple trees closest to the grounds of the Chase House. He predicted that these apple trees would have black streaks by May and June and exhibit signs of shepard's crook in July.

At least fifteen major diseases and insect species can attack and kill apple trees. The *Forgotten Fruits Manual* cautions us that some of the strains are getting more virulent and spreading to growing regions where they had previously been only minor problems: "For instance, fire blight was not considered a major apple disease in the Great Lakes region until the 1980s, but the number of episodes of fire blight have since reached epidemic proportions in that region, damaging or destroying many young trees."

We walked past the rare trees in this part of the orchard that had avoided the disease so far: an Allington Pippin and a Connell Red, a fragrant variety discovered in Wisconsin. When I bite into an Allington Pippin, a British variety from 1894, I find a hint of tropical fruit, mainly guava and pineapple. Mia pronounces it "brightly sweet." The apple grows sweeter as it ages in our kitchen, *en plein air*, acquiring a heady apple-juice flavor.

At the Chase Orchard, the undergrowth was rich with verbena, salvia, some milkweed, but also the invasive mule ears and wild lettuce—perhaps miner's lettuce with the milky sap. Gordon mentioned that the lettuce can be mashed into water and fermented in the sun and then sprayed on plants to discourage grasshoppers from eating the leaves. Just as the elder apple trees had been my guardians earlier that morning, I felt that Gordon was now transmitting plant knowledge as only an elder can.

Among the young apple trees, less than a decade old, were veterans over a century old with character to spare. On such trees, the main trunk was mostly all hollowed out, and only one bough, usually on the north side, had vigorous growth. These veterans still bear apples as I had witnessed during the fall harvest a couple of years back. But last year, the orchard had no apples.

"Last year was brutal," Dave Kenneke said.

A cold snap had run through the area. A killer snap. I paused alongside the maternal apple tree that died from it. Gordon reminisced about the cold snap in the 1970s when New Mexico lost several orchards. I nodded, recalling John Trujillo's orchard.

As the afternoon drew to a close, it felt that we were kindred spirits who had spent almost three days in close proximity and shared a way of thinking about apple trees. When Gordon said that every plant and animal has a spirit, Dave's face lit up. "That is what I was remarking to them yesterday," he said, gesturing toward his young grounds crew members, who smiled shyly.

"It's a female spirit," Gordon added. "Touch 'em!"

We rolled our eyes.

"Grab a twig and let go, you'll feel a cool streak."

All afternoon, the storm kept threatening to rush in. For half an hour, however, the gloom lifted and the sun lit up the stunning mountain ranges around us. Then the clouds came back even stronger, graying and chilling the air, announcing that the snowstorm was on its way.

Gordon left us with a parting thought: Be curious and have an investigative spirit. As we drove away, cattle rested under towering old Seckel pear trees next to the Chase Orchard. Driving through town, Gordon reminisced as we passed by some shuttered old buildings: "That used to be the Ranch Bar—a good place to see a fistfight. . . ." He was moving through his past in real time, in a way that I would never be able to do.

On the winding road from Cimarron to Taos, we passed the stretch from where Gordon had taken the Bear Tree scion after he found that the original tree wasn't in good shape. The scion for the Cimarron Canyon tree was taken near a state park, and the tree must have fallen since, for we didn't find it as we drove by. A little farther, massive rock faces juddered out of the earth, and the one jutting out on its own is called Devil's Mailbox. On the unceasing windy road, I was feeling queasy in the back of the truck but refused Gordon's offer of a "barf bag." The snow flurries had thickened, and the curving earth around us was shrouded in white. We drove past Peñasco to Truchas, where the morning's storm had soaked through, and the road was so muddy that even Gordon's truck swayed perilously. At his farm, when we spilled out of the truck, he said that he would drive me back to the road since Michael might not be able to clear the muddy stretch. It was snowing thickly, and just then we saw the silver Honda wavering toward us. Michael and the girls were coming in through the mud!

Somehow the Honda made it, while my own sneakers were coated with so much mud that I was clomping around instead of walking. I saw the beaming faces of the girls in the car and moved toward them, handing them back the Krishna stuffie, on loan from Pika. I must have looked like an apparition in swirling snow, and they gestured to me to get into the car. But I was waiting for the crabapple tree. Gordon's gift.

With knowing arms, he got the tree out of his truck and scooched it into the trunk of the Honda. And then we began our wobbly ride out into the snow-smacked evening.

"It is a French variety," I told Michael in the car. "*Muscat de Dieppe* on 890 rootstock. A cider apple, and it'll be good to cook with. The tree has a beautiful structure."

Michael nodded contentedly, and we went careening through the evening, through the surreal hillocks of clay that stretched on and would have you thinking that you're on a different planet, with an otherworldly landscape. Incredibly, people are born, rush through their years, and life passes them by. This is one way to slow down time; to prune an apple tree into a lovely structure, and care for it until it bears fruit for children who may love the apples someday.

Chapter Fourteen

The Generosity of Blossoms

APRIL

Las Golondrinas apple tree in bloom. Photo by Priyanka Kumar.

I

The swallows, las golondrinas, had arrived not even a week back. They darted across the bright April sky with the languid vigor of imminent mating. Early that morning, Sean Paloheimo had seen his first tanager of the season. A female. When I mentioned the tanager's preference for cut oranges, Sean smiled politely. "That's what my dad does. But I find that orange suet lasts longer." Blond hair spilled out of Sean's dark baseball cap and glinted in the sunlight. Michael and I had met him early in the morning at El Rancho de Las Golondrinas, where Sean has deep roots. The springtime sun would soon brighten, recalling the blazing summer days I've spent wandering in these grounds in years past.

We made our way to a small grove of historic apple trees in an area that isn't accessible to the public. The morning light breathed the softness of dusky clouds, and when a glorious pear tree came into view, I gasped upon seeing it in full bloom. Pear flowers are creamy white while apple flowers are white with a pink blush. The tree before me was exceptionally tall—the structure of a pear tree tends to be more upright when compared to the inverted-shrub branching structure of a cultivated apple tree. Despite the visual cues if you've ever confused a pear and an apple tree in the absence of telltale fruit, forgive yourself, and recall that Linnaeus himself was wrong about the relation between the two trees. In 1740, Philip Miller rightly grouped the cultivated apple and Malus sylvestris, as well as five other crabapple species, together in the genus Malus. Miller kept this group separate from the pear because the scion from one tree cannot be grafted onto the rootstalk of the other.

Twenty-seven years later, in 1767, Linnaeus disagreed and grouped the cultivated apple and the crabapple with the pear, as well as the quince, in the genus Pyrus, against the judgement of other botanists. Miller, who was also a fruit grower and the first to authoritatively characterize the genus Malus, begged to differ in his 1768 Gardener's Dictionary:

> Dr. Linnaeus has joined the Pear, Apple, and Quince together, making them all of the same ... and has reduced all the varieties of each to one species. ... But where the fruit is admitted as a distinguishing character of the genus, the apple should be separated from the pear, this distinction being found in nature; for their fruits will not take by budding or grafting each other, though it be performed with utmost care. ... Therefore I shall beg leave to continue the separation of the Apple from the Pear, as hath always been practiced by botanists of his time.

With this flourish, Miller defied the godly Linnaeus, extracted the misfit apple out of *Pyrus*, and moved it back into *Malus*.

I took time to relish the magnificence of the pear tree's thickly dotted, cool-white blooms, which was as well, for upon turning around I was met with the sight of a handful of ancient apple trees who had given up the ghost. On one dreary, stout trunk, I saw a faint set of teeth marks that had fatally grazed the bark.

"Our donkeys initially left them alone," Sean said. "But then we got two new donkeys who taught them to girdle the trees." He sighed. With his hands full as the head of operations and, now, assistant director, of the Living History Museum, he had neglected to check on the oldest apple trees until it was too late. The three historic apple trees flanking Sean's house remained protected for an obvious reason. He had kept an eye on them. We walked toward his house, passing two of the apple trees in gorgeous halfway bloom. Apple blossoms are fabled for their beauty and healing powers and, in astrology, are in the domain of Venus. The biologist David George Haskell quotes from the doctrine of signatures: "The showy, scented blossoms of the apple tree" can "heal disorders of fertility and complexion." While I haven't used apple blossoms medicinally, it looked as though Venus had sprinkled her enchanting beauty before me.

We came upon a modest pea-green pond facing the house. Twenty years back, Sean moved in here, into what was then an off-grid log cabin without plumbing; over the years, he attached a structure on each side and now lives in the house with his wife and two dogs. The dogs barked wildly from the portal as we approached, and Sean almost managed to calm them.

Crossing a small meadow, I approached an apple tree with a classically lovely structure. This must be the queen of all the historic apple trees left on the property. I was admiring the tree, in peak bloom, when Sean approached and told me that his beloved old dog is buried underneath.

"Now he feeds the tree," he said, a streak of emotion crossing his face.

Sean has known these apple trees since he was a tot, and he believes that they are well over ninety years old. When his great-grandmother Leonora Curtin bought the property in 1932, the trees were still standing. He told me that the variety is Wolf Willow, and it is also found in the Real Orchard. But there is no known variety by that name, and the apple's red coloration with a sandy crown and its characteristics—a generously sized, slightly tart baking apple that matures early in the season—all point to a Wisconsin variety called the Wolf River. Later when I checked in with apple historian Dan Bussey, he wrote back: "Your description is spot on for Wolf River." He added that "names and synonyms are fun to track down to know where and when they were used," and now he plans to add Wolf Willow to his growing list. Three brilliant specimens hew the grounds around Sean's house; the ancestral trees are fed by the small pond (which initially irrigated the trees) and the Tonquin acequia that runs through the property. Over the last fifteen years, however, the acequia has noticeably dried. Even in springtime, the flow was timid.

Having grown up here, Sean attested that the flow was once robust. "But now, in mid-summer, it's down to a trickle," he said quietly. In dry times, the Las Golondrinas staff relies on a well, which of course relies on groundwater. We began to overuse groundwater in the twentieth century; since the days when these apple trees were first planted, we've collectively drained far more groundwater than is our fair share, and now the precious source is being exhausted faster than the ground's ability to replenish it.

The old apple trees that have fed from the Tonquin acequia and survived are terrific bearers. Forty years back, Sean's grandmother

Leonora Curtin Jr. would ask the Santo Domingo Pueblo to send over a dump truck and have it filled with apples. Some twenty years back, the staff at Las Golondrinas began to think about the trees' mortality and how they might perpetuate the Wolf River variety; they decided to graft the historic trees onto fresh rootstock. The grafting was underway when Sean came back to the ranch after studying photography at the University of New Mexico, followed by a stint in Bend, Oregon, where he photographed snowboarders on Mount Bachelor and other mountains. When Sean returned home, he saw the Las Golondrinas' groundskeeper completing the last of the grafts. Seeing Julie Anna at work was a familiar sight; Sean had seen her puttering about in the fields and garden since he was four.

The sun gained in strength as we made our way from the historic apple grove to the grafted orchard. Along the path, a mature apple tree in bloom caught my eye. The tree was set against an old cabin and flanked by two carts, which looked beautiful though ragged with age.
  "The cartwheels are made of cottonwood," Sean said.
  Looking beyond the apple tree, I could see a line of cottonwoods running along the Tonquin acequia. While Sean has piped parts of the acequia to save water, he lets it run naturally where the cottonwoods grow, so that those trees will continue to flourish. While Sean and Michael conversed in the backdrop about drying acequias in a time of drought, I lingered with the apple tree by the cabin, watching a bumblebee pollinate the blooms. The delicate business of apple pollination doesn't always come off successfully. In April and May, it can get excessively windy in this part of the world—and wind is no ally of blooms. Apple pollen is too heavy for wind alone to carry, which is why pollinators from wasps to native bees and bumblebees are critical for apple trees. Bees and wasps are cousins, but bees feed their young pollen, not insects as wasps do. The bumblebee I was observing is a social creature and belongs to the largest family of bees, Apidae, which also includes honey bees. Whereas honey bees are commonly used and even rented to pollinate fruit trees, native bees such as solitary bees—including mason, sweat, mining, and polyester bees—play an underappreciated role in orchards. Along with bumblebees, solitary bees dominate

the diversity of bees in apple orchards. Solitary bees are actually more effective at pollinating apple blossoms on a per-visit basis compared to the honey bee: They exhibit a greater preference for apple blossoms, carry more pollen, and have a higher rate of setting fruit. Incredibly, they also have a higher tolerance for harsh weather conditions, such as heat and aridity, and will fly even when it is raining—all of which may make them a climate-friendly bee. Solitary bees are so beneficial that they are believed to account for more than 50 percent of apple production, based on studies conducted at orchards in Ontario, New York, North Carolina, Pennsylvania, and Wisconsin. Apple orchardists have their own preferences—those in New York state rely heavily on solitary bees, whereas the big apple producers in Washington state depend heavily on honey bees. It is worth noting that solitary bees can be more susceptible to the toxic effects of pesticides due to their ground-nesting habit, whereas honey bees live in large colonies and have mechanisms for purifying their hives.

Wild apples, including *Malus sieversii*, *Malus sylvestris*, and *Malus orientalis*, are pollinated not only by bees but also insects from the Syrphidae family of flies. With their yellow-striped black abdomens, Syrphids naively look like wasps—which may be a case of evolutionary mimicry. Syrphids produce larvae that eat aphids, so wild apple trees get two benefits from their association with syrphids.

During roughly ten days when apple flowers are in bloom, they need at least a single tranquil day for pollinators to carry their pollen to a genetically compatible variety and cross-pollinate the trees. So, apple trees depend on the generosity of wasps and bees. As in many plant families, including several species of Rosaceae, after the apple flower is fertilized, the petals fall off but the sepals remain attached throughout fruit development, eventually drying up.

The sizeable bumblebee, all yellow and black, landed on a cloudy cluster of blooms and drank its fill before moving on to an adjacent cluster. I drew closer to the blooms, inhaling their spring-sweet aroma. Emily Dickinson once wrote to her aunt: "The lovely flowers embarrass me. They make me regret I am not a Bee." Absorbed in lotus-pink buds and milky-white flowers as a bumblebee careened from one flower to the next, I felt like I was becoming drunk with spring, like cedar waxwings on cherries, and coyotes on the Wolf River apples by Sean's house.

"They just fall asleep in the meadow here, they get so drunk," he'd told me. "And my dogs bark and bark at them. And I say to the coyotes: C'mon guys, go sleep somewhere else!"

Once a year, an adult bear lumbers by the historic apple trees, journeying here from the Jemez Mountains. "He knows the place is full of apples," Sean says, shaking his head, eyes sparkling. This particular bear has enjoyed devouring the Wolf River apples for many, many years, but his reverie ended recently when he got into the neighbor's beehives.

"They called Fish and Wildlife, who came and relocated the bear," Sean said.

Chasing the beauty of blossoms, I found it thrilling to commune with apple trees—each exceptionally unique—that Spanish ranchers had likely planted here almost a century back. Did they mean to plant the trees roughly near the path that leads to the pumpkin and bean field, so that harvesters might have some refreshment while working in the field? The Wolf River ripens early, and the fruit is unusually large. Though a somewhat tart apple, it must have been a bracing snack for thirsty farmworkers on hot August days. The past is present here: The apple trees that remain are still suckling from the same acequia that has been flowing here since 1710. As we walked toward the grafted orchard, I recalled how Sean's great-grandmother Leonora Curtin Sr. also left a wetland for all of us to wander through. Each autumn when I am ready to sink into golden heart-shaped leaves, I pilgrimage through the ancient cottonwoods in the wetland. It is a true act of generosity to share beauty with those whom you will never know.

The grafted orchard, set out along a grid, compared unfavorably with the beauty of the irregularly planted old trees. Being among the historic trees had felt right—it was the wildness that I loved.

At the grafted orchard, even as I admired Julie Anna's handiwork and the trees' sturdy branches and structure that tended outward, I remained at the edge. I didn't feel the urge to walk through the grid, which is undoubtedly an efficient system to harvest fruit. In the ancient grove, the beauty of the flowering trees, vying to offer their opalescent white flowers to the sky—and even the dark cacao beauty of the dead trees—had entered my heart.

The apple tree by Sean's house was exceedingly tall, and two mourning doves, spring mates, had flown to the top and lingered, trying out high boughs and conferring before flying off with a thrumming flutter. Not only humans but also birds and a beloved dog take shelter under that tree. Whereas the trees I now stood in front of would certainly give good fruit but not any shelter. To be sure, they were planted with excellent intentions on rootstock that is widely used today. There are sensible reasons to use dwarf rootstock: People don't have to wait as long for trees to bear fruit; dwarf trees are more contained and are easier to prune and harvest. But the magic of the old trees uplifts the heart.

II

We parted ways with Sean under a swallow-filled sky. Michael and I walked over to a somewhat hidden pond on the property, rimmed with overgrown brush, where we observed birds. Suddenly, we were drenched in the resonant sounds of *Agelaius phoeniceus*, red-winged blackbirds; in fresh plumage, the birds flitted among the leafy willows that hewed the water. Michael recalled how he'd hear red-winged blackbirds by a pond in his neighborhood when he was in second grade.

"I also collected tadpoles in that pond," he said. "And skated on it in the winter."

As we spoke, I noted plump tadpoles swimming underwater in the pond before us. Soon, we saw a Wilson's warbler, our first this spring. Seeing spring's first warblers is a shot of rasa, a pathway to becoming spring. I had seen my first thrush of the season that very morning—a hermit thrush who'd raced fetchingly past our garden.

Michael spoke about the pond he had loved when he was six or seven, where tadpoles swam and red-winged blackbirds charged the air with trilling calls. That magical pond soon got paved over. "I was shocked one day when I saw the townhomes they had put there instead," he said. "It was all concrete. I biked along the street and felt like I was in a canyon, hemmed in by the walls of the three-story-high townhouses. I felt hollow."

The old pond is partly where Michael's love for the natural world was seeded. For a boy on a bike, a pond is more than simply a pond; it is a retreat, a refuge, a space to exercise curiosity—and it vanished all too soon. The sacred continues to be paved over for the mundane.

Decades later, it felt meaningful to sit together on a log bench in Las Golondrinas, by the hidden pond. It has taken me years to realize that this pond was here, and I rued that I had walked past dozens of times without being any wiser. At least for this morning, we had this watery refuge to ourselves. Fortunately, Leonora Curtin Sr. and Jr., and their families, had the vision to be caretakers of this pond and a historic grove of apple trees. Interestingly, the Wolf River apple was established in 1856 along Wisconsin's Wolf River by a woodsman from Québec, the same province Michael's maternal family had migrated to in the 1600s. When I experience how old apples bring history to life, I feel sure that the practice of being a caretaker is essential in a time when land that isn't protected is inexorably earmarked for development. The problem of habitat loss caused by development is, of course, a global issue. Over a decade back, the Kazakh scientist Dzhangaliev wrote to his government, cautioning that less than 30 percent of the country's original wild stands of apple trees remain. "Do you want to destroy the tree shown on the national emblem of Kazakhstan?" he asked.

Just as the surviving old apple groves in America could be seen as a national treasure, the Kazakh apple forests are a planetary treasure and a critical genetic seedbank. Researchers and travelers to those forests have remarked on the diversity of herbs—tarragon, oregano, thyme, mint—and flowering plants in the shrubby understory of the wild fruit trees. The forest canopy mainly consists, remarkably, of fruit trees—apples, but also apricots and small pears—with some other trees such as elm and maple sprinkled in.

In addition to their beauty and cultural importance, there are scientific reasons to preserve these ancient forests. Today, scientists are using *Malus sieversii* to determine how cultivar apples could become more disease- and pest resistant. Horticulturalists have discovered that some populations of Kazakh trees, such as those in the cold and high mountainous regions, show significant resistance to apple scab—the most important fungal disease of apples—as well as to fire blight, among other devastating diseases. Interestingly, research indicates that the apple scab fungus also originated in central Asia and that *Malus sieversii* was its primary host, and *Malus domestica* and *Malus sylvestris* secondary hosts. In fact, the woodland crabapple only became infected with this fungus after contact with *Malus domestica*, which independently supports the hypothesis that *Malus sieversii* was the original progenitor of the cultivar apple.

The past and the future are strangely intertwined. To brace ourselves for the future, we must now reach back to our primeval past. "Locked away in the genetic codes [of the apples in the ancient Asian forests] are the still unexplored possibilities of what an apple might become," writes author Frank Browning. But these unexplored possibilities are dwindling as biodiversity diminishes. I was eager to explore some of these possibilities in the Tian Shan mountains, but I was kept busy navigating Kazakh bureaucracy. After a flurry of correspondence, I at last applied for a permit to visit the wild apple forests.

Michael and I lingered in the shade, gazing at light glinting in pond water. A black phoebe alighted on a lone reed swaying in the breeze, and the phoebe also chose to linger. When the time came to leave, the scent of apple blossoms was still in me, and I felt that I had hovered around the bird-rich pond in the same way as the bumblebee had lingered wantonly among apple blooms.

Recollections may vanish or turn into half-memories, but powerful remembrances sometimes return to embrace us. My apple love connects me to the past and influences my future; I now feel inspired to create apple rasa for others. At Las Golondrinas, I saw the same trees, only they are fuller and taller, as those Leonora Curtin saw in 1932. In that year, my parents weren't born yet, but these trees were already vibrant and strong. Astonishingly, decked out like debutantes, they still sing their flowery songs—now, when my own children are in elementary school.

Perhaps a Spanish rancher planted the trees a hundred years back with the same fervent hope with which Michael and I have planted and care for our young apple trees. Considering the long history that we share with apple trees sustains my enthusiasm to go on growing them. It makes me wonder: If our trees thrive, who will enjoy their blooms a hundred years from now? Planting trees is an act of generosity, a way of communing with those who will come after us. The act of growing fruit trees asks that we change our relationship with time. "No great thing is created suddenly; any more than a bunch of grapes or a fig," Epictetus wrote. "If you tell me, that you desire a fig, I answer you, that there must be time. Let it first blossom, then bear fruit, then ripen." Those who will one day enjoy the trees we tend today may not know our names, but in some future autumn, they will harvest the ripe fruits of our handiwork and, in a future spring, animated by las golondrinas, they will know us by the scent of opalescent blossoms.

Chapter Fifteen
Cow Creek Runs Through Paradise

MAY

Pear tree in bloom. Photo by Priyanka Kumar.

I

Marcelino's tan complements his sea-gray eyes. He is known to be shy and, when he started walking me through his orchard, he was fairly quiet. But his marine eyes are big and open, like his golden laugh. Once Marcy (as he's called) befriends you, his face glows with musical laughter. He has the air of an Italian orchardist who is content with life and couldn't imagine another way of living. A bachelor in his early sixties, Marcy lives in a village of no more than eighty people, most of whom are over sixty. You may have heard of some place where time has stopped, but when you experience such a village, time stills within you. You might even wonder why you've been running around. At Marcy's orchard, I discovered something indefinable that I had been searching for, a little over an hour away from home.

"If there's a kid here, I haven't seen him," said Steve, who had brought me here, referring to the village's aging population. A contractor who restores wetlands with heavy equipment, Steve was hired a couple of years back by the Upper Pecos Watershed Association and some government agencies to restore Cow Creek—which is how he came to know Marcy and some others in this halcyon village in the Pecos area. Over a year back, Steve and I ran into each other at Gordon Tooley's grafting workshop; Steve's son and Pika, both eight, have attended the same schools since the age of four.

Before entering the village, we drove past three ranches owned by "outsiders"—Val Kilmer owned one ranch before selling it to a Texas oilman. But the village proper has stayed true to its Spanish origins and is knit together with long-standing familial bonds. On the Wednesday in mid-May when Steve and I visited Marcy's orchard and inspected some sections of Cow Creek, we would also run into Marcy's brother Elias in his vegetable field, and another brother, James, driving into the village in his jeep. If you have any business here, as Steve does, you simply stop and talk it out. Steve was trying to track down the Armijo brothers; their old man died over a year back, and the brothers, who live in Albuquerque, haven't paid property taxes for the village land they inherited. Steve isn't out to collect taxes, but he needs their permission to restore the part of Cow Creek that runs through the Armijo property. He had been given an Albuquerque address for the brothers and attempted to track them down but found the house to be abandoned.

As we walked through Marcy's orchard, Steve asked him about the Armijo boys.

"They're not country boys," Marcy said. "I haven't seen them since the father died. He didn't work the land either. The boys just came here in the summers to fish. They are city boys."

Marcy, au contraire, is all but married to the land. Steve remarked to me that whenever he stops by the orchard, Marcy is always on site. This time, Steve had knocked at two low adobe structures, which look ancient, with battered furniture outside (Marcy is a carpenter), but no one answered. A few minutes later, we saw a shadowy figure cross the orchard. Marcy. Despite Steve's warning that Marcy is shy, soon after we met, he clarified my name. "Priyanka? I have a niece called Bianca." He found the coincidence amusing.

"Where are you from?" was one of his first questions.

"From Santa Fe." After a pause, I sensed that Marcy wasn't convinced by my answer. And I briefly described my childhood in the foothills of the Himalayas.

Intrigued, he questioned me closely about the fruit grown there.

I told him that I grew up among apple trees flourishing on mountainsides and in valleys. This changed everything. He looked at me as a kindred spirit, and I was allowed to enter his universe. "Do they grow fruit other than apples in Himachal Pradesh?" he asked.

"They grow apricots, as well."

Thanks to Marcy's caretaking, the apple grove that his grandfather planted in 1959 has grown into a paradise. The understory is lush and grassy, and the trees—mostly apple, but also pear and cherry—tower over you. Over thirty mature apple trees have survived, and Marcy told me that more than a few were planted before 1959. He pointed out a Jonathan Red, one of his favorites—and his face broke into a juicy smile. Over a hundred years old, the Jonathan variety is an offspring of the Esopus Spitzenburg. The midsize apple is shaped like a child's ball that is slightly flattened; it has a spicy, sweet-tart flavor and is seen as the quintessential American apple. Jonathan trees are strong bearers, and the King Red variety is used to make candied apples.

"I lost the apple tree right next to it because of the drought," Marcy added ruefully, referring to the time a handful of years back when the drought was painful and there was no winter precipitation and little snowmelt.

It helps that Cow Creek runs through the property and, when there is good precipitation, Marcy diverts the water into ditches and flood-irrigates the orchard. One of the ditches runs along the orchard's edge, against a fence line with his neighbor. The fence line may well be dividing two vegetation zones. While the neighbor's field clearly looks like it is suffering through a megadrought, the lushness of Marcy's land is nothing short of astonishing. Marcy's mature trees keep his property shaded and cool, and grasses flow abundantly. Steve told me that in the summer, the place brims with the loveliest flowers. Not only the neighbor's property, but also others that I later walked through in the area look parched and desertified—though those owners presumably have similar water rights as Marcy.

"A couple weeks back, all these trees were in bloom," he said with a beatific smile, arms involuntarily gesturing up to the mature fruit trees.

I eyed the trees, some still blushing pink. The blooms were effusive, as though inviting one to spread a mat under the pinkly green canopies. As I examined the blooms, Marcy didn't interrupt me. Clearly, he took real pleasure in them, and his delight was infectious. I could all but see myself wandering once again among apple trees as a child. My aunt's words returned to me: "You always had an apple in your hand." Holding on to an apple meant more than having a snack at hand: It meant an aromatic palm, and hewing to a way of life that Marcy still lives. It seemed to me that Marcy was nearest to himself while wandering among the trees—just like I was as a child.

"Do you prune the trees?" I asked.

"I do. But I haven't in the last couple of years." He chuckled. "That's why they're so thick over here." He gestured to some branches crowding in on each other.

The canopy may have been beautiful but wouldn't pass muster with Gordon Tooley. When I asked Marcy about the apple varieties, he said, "I don't know their names, but I would like to." I offered to return sometime to take samples and get them identified at a lab. It would be a way to return Marcy's gift—he had welcomed me, a stranger, into his paradisical orchard.

Marcy nodded gladly. "I would like to know," he repeated.

Steve told me that Marcy's trees are terrific bearers. I lingered, taking a last look at the blooms before following him out of the main orchard. "What do you do with the apples?" I asked.

He looked back and gave me a sidelong glance. "I share them with the bears," his face bursting into a smile.

Experiencing Marcy's orchard truly felt like being with a kindred spirit. My heart was humming with joy. He shares his apples with bears! Why not? Among his apple trees is a green apple, bland and a bit mushy, but the bears readily eat it. So, he has let the tree be. Marcy also shares the orchard's apples with his extended family and takes some to his brother's farm for the goats, who relish the fruit. Connected to his trees and wild animals, Marcy intuitively takes care of the biodiversity on his land. This is why his understory is singing, as are his trees. The bears in turn have taken care of him, as I would soon discover.

II

The apple trees in Marcy's orchard were largely grown from seedlings, and some from root suckers taken from a mother tree, likely a Winesap, that grows in an incredibly remote part of a village. Two years back, Marcy and Steve drove on inaccessible dirt roads to find the tree, which was still living. I craved to see the mother tree, but it was too early to ask. I suspect one has to earn trust around here—where people have known each other since they were babies—before inquiring about a treasure like the mother Winesap.

In the orchard, horse manure thickly dotted the grass. Here and there, Marcy had gathered some manure and shored it up against the base of apple trees. I asked if he owns horses. He laughed enigmatically before admitting that he owns a filly and a mare. Later, Steve told me that a few horses run freely in the village; perhaps they are somewhat wild or communally owned. Whoever needs to use a horse will go ahead and harness the animal. When I remarked on the flavor of communal living in the village, Steve added, half-jokingly: "But if you cross someone the wrong way, they'll shoot you."

Marcy seemed to be such a joyful spirit that I couldn't imagine him shooting anyone. Steve later told me that some of Marcy's twelve siblings consider his values to be a little odd.

As we walked toward the wild part of the orchard, by the Alamosa cliffs, Marcy said that two of his brothers wanted to cut down a pair of towering oaks that grow against the cliffs on his property. But Marcy demurred. "I didn't want them cutting down those oaks or the oak near my house," he said. A little later, when we were studying his wild apple trees, he added, "The ground underneath the oaks is really black," his hand involuntarily moving as though sifting through rich dark soil. Soon, Marcy was asking me about the fruit in the eastern foothills of the Himalayas where I had also lived in as a child. I told him about the tropical fruits we grew in Assam. "Have you tasted jackfruit?" I asked.

"I have heard of it but never tasted it."

I described the taste of ripe jackfruit bulbs, like banana and songs of pineapple, guava, and mango all rolled into a deliciously soft fruit, and how a monstrous jackfruit tree was the star of our backyard. And how we used banana leaves as plates during outdoor feasts.

We were now in his wild orchard, and I was struck with awe when I saw the gift of the bears. The apple-satiated bears had planted this orchard for Marcy. The orchard abuts the Alamosa cliffs, which are sheer and tawny beige and striated with limestone. The magnificent beauty of the towering apple trees, some still in bloom, against the sheer cliffside cannot be overstated. The thirteen feral apple trees are punctuated by the pair of surviving oaks and some mature box elders. And in this heavenly line of trees grows the apple tree that Steve calls "the true Marcelino."

Two years back, when Steve was picking apples from this tree, his wetland crew asked him to stop. They were camping along the creek for a couple of months while restoring it, and each night after dinner, the crew would stew these apples, no sugar needed, for dessert.

"It's the sweetest apple I've ever tasted," Steve said. "The best I've ever tasted."

High praise from a man who gets around a lot, restoring wetlands from one end of the state to another, and tastes apples (and picks mushrooms) wherever he can find them. "When I first came here, I loved it," Steve said about Marcy's orchard. "You can just wander around tasting different fruit. Some of the apples are spitters, but I bet they would be great for cider." He plans to borrow a cider press and set it up here one day for Marcy and his extended family to use.

Marcy has made a real impression on Steve, and now he is also giving me a serious dose of joy. Seeing root suckers coming out of the true Marcelino, Steve asked if we might shovel some out. Marcy readily agreed and brought back a shovel he had forged himself—heavy on the top and bottom for ease of use—and six old planter pots. We all pitched in, shoveling and gathering the suckers, which looked like saplings, into the pots.

Malus trees send out root suckers, and in central Asia ancestral people would have transplanted new saplings from trees bearing "good" apples from the forest to their gardens. If enough trees carrying maternal *Malus sylvestris* ancestry were introduced to cultivated heterogenous apple trees in this way, and cross-pollination left unfettered, this is another way *Malus domestica* could have acquired a substantial portion of the *Malus sylvestris* chloroplast genome. A team of scientists led by Svetlana Nikiforova outlined this alternate scenario for how *Malus sylvestris*, the European crabapple, became the dominant contributor to the chloroplast genome of the cultivated apple. The scientists caution, however, that while many *Malus domestica* chloroplast genomes appear to have been acquired from *Malus sylvestris*, that is not the case for all cultivated apples. The genetic history of the apple is layered with complexity, and it felt thrilling to engage in a practice that ancestral Kazakh people had likely used to propagate apple trees.

We carried our potted suckers back by hand and in a wheelbarrow. Steve planned to plant five in his five-acre farm in Espanola, and Michael and I would plant one in our small but swelling orchard.

When Steve mentioned that I would one day go to Kazakhstan, Marcy was fascinated. The sun shining on his face and blue-gray eyes, he said, "You can't get bored—there are so many varieties of apples out there. Just thinking about it is exciting."

He asked that I bring him some seeds from Kazakhstan (later, Michael told me that might be illegal). If not seeds, I would love to bring back something for Marcy that won't get me into trouble at customs.

Marcy and his orchard blend into each other; they are made of the same cloth. There is an inner wildness in Marcy that manifests in his orchard. I gasped when I saw the wild trees that the bears have planted here: a line of thirteen trees against the feral Alamosa cliffs. These bear trees are nestled into the wild cliffs. It is as though Marcy lives in a Shangri La of his own making—in concert with his ancestors, the bears, flickers, oaks, and the mother Winesap. The beauty of the place swallowed me up. I understood why Marcy glows, why his laugh is musical. Of course, some might hear sinister notes, imagining bears lurking behind trees (they live on the other side of the cliffs) or snakes crisscrossing the abundant grasses. But once you tune in to the music of the orchard, with Marcy as the wild conductor, your heart swells with delight.

III

After surveying the wild orchard, we took a walk along the azure artery that runs through the land; the water in Cow Creek was surprisingly blue, and we surveyed the restoration work that Steve and his team had done, such as dropping boulders at strategic locations to slow down the waterflow. Marcy paused to pick up a cottonwood leaf that had fallen on the bank. "Cottonwood leaves take a long time to break down," he said, showing me the tawny leaf, which was perfectly intact from last fall when it had fallen to the ground.

A huge wildfire—the Calf Canyon Fire—ran through these parts not long back, and Round Mountain, one of the peaks that frames this valley, still wears charred skin. When we had driven into the village and crossed the Pecos River, Steve was relieved to see that the water looked clearer than he had expected—it was no longer sullied with wildfire silt, ash, or other debris. Marcy lives among persistent threats of wildfires, drought, and poverty, but it struck me that he is one of the richest men for miles around. He is rich in trees, beauty, and the unique love he has cultivated for the land and its creatures. In my years of wandering through orchards, I have scarcely found a happier or a more contented man.

After packing up the true Marcelinos in the trunk of Steve's jeep, we said goodbyes, and Steve drove me to a couple of spots where we walked along the river to see how his restoration work was coming along. The adjoining property at the edge of the village abutted the wild, even more so than Marcy's. Beavers had eaten and finished off some young cottonwoods that had been planted, and some large stones in the creek had shifted position, but the overall flow of the creek had been slowed and its banks graded so that the water would irrigate the surrounding willows rather than rushing through as though in a gully. In these willows, I saw and heard warblers; a yellow flew across the creek to alight on a stem, a Wilson's hopped from one willow to the next, and a yellow-breasted chat scented the air with varied song.

Later, Steve and I drove through the village, which sits in a hidden valley, and the feeling persisted that if I were to overlook the occasional automobile or rusting farm machinery, time has bypassed this place. We drove to the northern section of Cow Creek, through a path that would have annihilated most vehicles. We began walking toward Cow Creek and startled a jackrabbit. Steve was looking for any signs of the Armijo boys and the location of an old beaver dam. There were some remnants of ancient houses, a couple of small two- or three-room structures that were reduced to striking ruins. No human in sight. We hiked over to the creek. The banks were thickly overgrown with pines and willows. The vegetation, dry brush, and piney branches tangled my arms, scratched and pulled, but at last we came to the clearing Steve was looking for. An old beaver dam, gorgeously built with willow and other boughs. Working for free, beavers do the necessary work of slowing down the river. In another part of the creek on Armijo property, the creek was entrenched six feet deep. Similarly, streams across the country have become degraded after the extirpation of beavers and wolves along with three centuries of grazing pressures; their ecosystem is simplified, and many streams no longer have good structure.

A week later I would encounter Reid Whittlesey, restoration director at the nonprofit Rio Grande Return. Backed by peer-reviewed scientific research, Rio Grande Return does "process-based" restoration that mimics natural processes to restore creeks, transforming them over time to wetland and riparian areas for a fraction of the cost. Done mostly by hand, this work involves planting willows and building BDAs, or beaver dam analogues, that reduce erosion.

The beaver dam on Cow Creek looked to be very old. "Are there any newer ones?" I asked Steve.

He frowned. "They shoot beavers around here.... Beavers like to eat plants. So, in that sense they are in competition with humans."

Some farmers dislike beavers because they believe that in a flash flood event beaver dams can cause flooding, for instance in areas across roads. On the other hand, in-stream structures also slow the flow of water, potentially reducing the negative impact of heavy rapid downpours. Later, Reid would affirm the embedded cultural animosity and institutional ignorance against the beaver, citing examples of US Fish and Wildlife game wardens who quietly look the other way when landowners shoot beavers.

The dams that beavers make bring back riparian and woody vegetation alongside streams, filter the water, and create the right conditions for cutthroat trout to make a comeback. All this beaver work ultimately creates havens of green space that serve not only as wildlife habitat but also as natural firebreaks. The latest studies show that areas where beavers had restored creeks to conditions approaching the historic norm did not burn down when a severe wildfire went through. "If we'd had beavers in all the watersheds in northern New Mexico, when the Hermit Peak and Calf Canyon fires went through, the Las Vegas water supply wouldn't have been endangered," Reid said.

Among the roots that beavers eat, however, are those of apple trees.

On our way out of the village, we came upon Marcy's brother Elias working his fields of corn and beans. Steve hopped out of the jeep to greet him and ask if he knew the whereabouts of the Armijo brothers.

"Our grandmothers were sisters," Elias said. "So, I should know, but I don't. What I heard is that they are in the city—their mother was from Albuquerque—and they are into drugs, they are partiers, some of them have even died...."

This was sobering news. We went on driving out of the valley, and just as we exited the village, we saw James, also Marcy's brother, driving into the village.

Steve idled the jeep to ask James a question. Would the Armijo brothers mind if he restored the creek that runs past their land?

"Just do a little," James said. "They won't care."

As though it were preordained, when Steve had just driven out of the village and we were approaching a high point with a stunning view of the valley below—framed by the snow-peaked Truchas Mountains—we saw Marcy's pickup. He was driving out, likely headed to a carpentry job—fitting custom-made windows—which he had mentioned earlier. Steve stopped, and I told Marcy that I'd seen two warblers, the yellow and the Wilson's, in the property adjoining his, and many other lovely birds, including the white-winged dove.

"That's why I didn't cut down the oaks," he said with a sparkle in his eye.

"You get it," I exclaimed. "It's all connected."

"I feel the connection," he said, sharing a smile of kinship before we parted ways.

When I returned home, the experience of feeling Marcy's orchard, and the connections it had fostered, made me consider how I spend my hours. To a lesser or greater degree, technology has swallowed us, and we all are familiar with the sight of people somberly going about their day while staring at phones. It is uncommon in everyday urban life to encounter a person with joy bubbling up from deep within. To be sure, I asked myself if I had simply imagined Marcy's joy. Then, toward the end of the day, I sat in a café to set down my experience in a notebook, and the contrast felt surreal. A few people walked by. By some coincidence, they happened to be Marcy's age; they looked burdened by life. One man seemed physically hunched over.

I woke up early the next morning, thinking about Marcy and his trees. It occurred to me that if he's in his mid-sixties, he was born not long after his grandfather planted the main orchard in 1959. When he was born, the trees must have been at most four years old. So, in a sense, he grew up with the trees. They have been his companions all his life. I understood why he spoke somberly about the loss of one particular apple tree to drought, why he opposed cutting down the oaks, and why he goes on planting flowers around the river. He has a living relationship with his fruit trees. When the tree roots send up nutrients to the fruit, they also feed Marcy, and the rasa that bubbles out of him is deeply infectious. After experiencing the way that Marcy walks through his orchard, something broke open—my understanding of how much rasa we can hold in our everyday lives. This rasa is catalyzed by nature, solitude, and community. I realized that I no longer wanted to "fit in" taking care of my fruit trees. In reality, the trees and the garden are taking care of me. I resolved to live more closely with my trees and allow them to claim me.

Chapter Sixteen

True Wild

JUNE

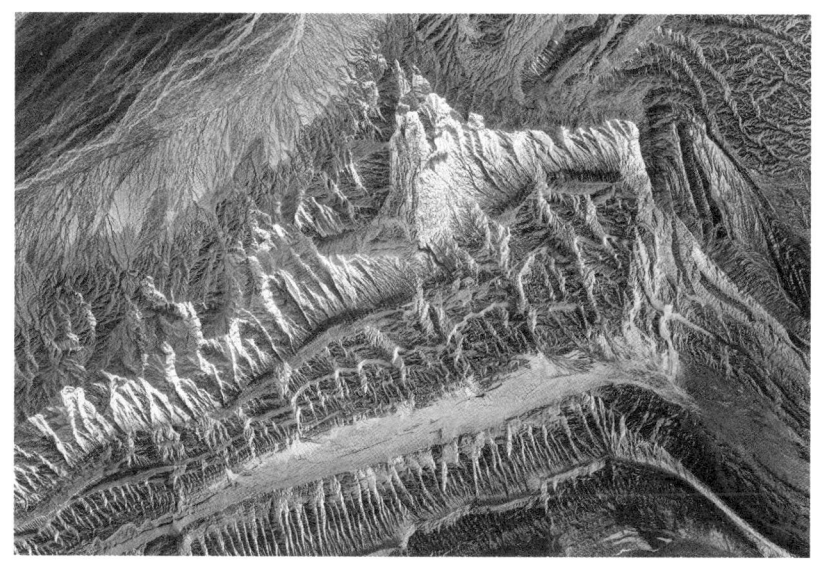

Tian Shan mountains from space. NASA photograph.

The desire to go to Kazakhstan was ticking in me. As much as I appreciate strips of wilderness, I also feel constrained and yearn for the true wild—to the extent that such places still exist. That fruit forests exist feels like a marvel, and I wanted a closer understanding of how wild apples differ from cultivars. Dzhangaliev had noted that wild apples grow under a wide range of climatic conditions—temperature, rainfall, humidity, dryness—in many types of soil, and under varying ecological conditions alongside other tree and shrub species. All this diversity is reflected in the wondrous polymorphisms (literally, "many forms") across the apple population, with the diversity concentrated at elevations from 2,600 to 6,600 feet in the fruit forests throughout the mountainous regions of Kazakhstan. I wanted to intimately experience the differences in wild apples selected from a range of locations, with their substantial biodiversity in terms of peel color, flesh color, spotting, juiciness, flavor, size and shape, and durability.

But what I read and the images I saw pointed to the degradation of even these ancient fruit forests. As Dzhangaliev warned, the land they need to thrive is diminishing, and many wild apple groves have already crumbled under the pressures of development around Almaty. A professor who'd stopped in Almaty a dozen years back said: "If you think there are places in the US that are turning suburban too fast, you should take a look at Almaty!" I wasn't too interested in seeing the green scraps left in the foothills of Almaty (named after the Kazakh word *Alma*, which means apple). I rued that the problem of habitat loss is universal; even the European crabapple, whose love story is written all over the cultivar apple we eat, is endangered now. Spengler, the archaeobotanist, had traveled to Kazakhstan over a decade back and gave me advice on how to prepare for my trip (bring sunscreen—it's hard to find it in central Asia). He also told me where I might see some apple trees near Almaty: "The apples are scattered in patches in the foothills around the city—mostly, they are short-growing trees or large bushes mixed with thorn bushes and tall grass." He cautioned, "There are lots of Russian-style dachas throughout the hills, so it does not feel remote or necessarily romantic, but it is still cool to see them."

I now wondered even more intently about the wild apple forests that were left intact in what is now a national park in Kazakhstan. When I inquired, I was once again told that I would need to follow an approved three-day itinerary to visit "genetic reservations" in the Zhongar-Alatau National Park. Essentially, the park administration had preapproved a cookie-cutter trip, and no further questions were encouraged. I clarified that I wanted to explore the wild apple forests and made a formal request: After the prescribed trip, could I spend three or four more days in the fruit forests, as well? And I waited to hear back.

All June, I expected to get an approval and prepared methodically for the trip. I considered taking the family along but felt daunted by the list of recommended vaccines, including for tick-borne encephalitis. Pika had never traveled out of the country and, in any case, we applied for her passport. I looked into flights and studied maps (in Russian), and it grew clear that the research trip would involve thirty long hours of flying and more than as many hours of driving to get to the national park. All for three days of tree time? It was all too fast and too little for my taste. I hoped that it would turn out to be a more expansive trip. As someone who has eschewed international travel for a decade, I also mulled over the carbon costs. I made arrangements to hire a translator and a driver. Spengler had told me that it would be all but impossible to get out to the wild forests without a driver. "There used to be a bus (Marshrutka) from the city to Medau in the mountains, there were some hiking trails along the way," he wrote, "but the last time I took that bus I was in graduate school, and I cannot remember the number."

I checked in regularly with my translator, and she assured me that my inquiry had been forwarded to the national park administration for approval.

Amid the uncertainty, I drew inspiration from Dzhangaliev's generosity in being a scientific caretaker of *Malus sieversii* and tried to better understand his remarks on the incredible variety within even the modern-era apple forests in Kazakhstan. In these forests, tree phenotypes differ "sharply" in terms of winter hardiness, immunity to diseases, and biochemicals in the fruit. Incredibly, some varieties of

Malus sieversii have seven times more vitamin C than cultivars, along with other nutrients such as folic acid (B9) and flavonoids that are not even found in Kazakh cultivars. Is this variability in wild apples due to underlying genetic differences between populations or the differences in abiotic and biotic conditions, such as soil, climate, and plant communities? This is no idle theoretical question. If these differences had a genetic origin, that would have major practical implications (in pharmaceuticals and nutrition, to begin with). But Dzhangaliev and his colleagues found it impossible to come up with answers based on field observations and, in the end, designed an ingenious set of field experiments: They grafted a selection of different scions onto wild apple seedling rootstock and grew them under experimental conditions.

What they discovered gives us a glimpse into the kaleidoscopic world of biodiversity: The daughter tree, and subsequent generations that were vegetatively propagated, had inherited many of the physical traits of the mother tree, the scion donor. Many of the differences in the wild form were preserved under cultivation. Daughter trees preserved winter hardiness, resistance to scab and powdery mildew, and did not suffer from winter drought. Small fruit remained small, and big fruit tended to remain big. Other properties, namely "color, form, flavor, storage life, transportability of fruits" were also preserved. So, many of the differences observed in the field appear to be genetic in origin! The diversity we see in cultivar apples is amazingly equaled or exceeded by that found in *Malus sieversii* and is genetically inherited.

Still, how did apples with varying sizes and tastes, many of them small and tart (*Malus sieversii*), become uniformly larger and mostly sweet (*Malus domestica*)? Fruit flavor refers to the balance between sugars and acid in the fruit, and both sweetness and acidity seem to have been modified as the apple was domesticated.

After its initial domestication from *Malus sieversii* in Kazakhstan, the "old" domestic apple spread west along the Silk Route from central Asia to Europe. Here, it hybridized with *Malus sylvestris*, the European crabapple, which has firm, tart, *bitter-sweet*, and *small* fruits. And, lo, a "new" *Malus domestica* emerged from this hybridization with firm, *sweet*, tart, and *larger* fruits.

The selection for fruit firmness was a significant advance—it not only helped with preservation of fruit but also meant the difference between crispy versus mealy texture. And humans tend to prefer crispy. Crispness or firmness seems to have been inherited mainly from *Malus sylvestris*. The introgression of genes—the bidirectional gene flow—was so robust that scientists now agree that *Malus domestica* is genetically as closely related to *Malus sylvestris* than to its original ancestor, *Malus sieversii*. Analysis of chloroplast DNA from *Malus domestica*, *Malus sylvestris*, and *Malus sieversii* has confirmed that many apple cultivars share more chloroplast DNA in common with the European crabapple than with the wild apple.

This incredible discovery illuminates that the apple is a transnational fruit, and the sweetness we distill today traces its origin to wild and even bitter ancestors. In retrospect, it makes sense that the apples we eat are close kin of the European crabapple. Some of the first North American apple trees came from grafted clones of European cultivars or their seeds. Scientists believe that these seeds carried new and unnamed genetic combinations likely chosen from desirable European mother cultivars. The wild crabapples of North America are *Malus fusca*, *Malus coronaria*, *Malus angustifolia*, and *Malus ioensis* (the one that died in my garden). They played no role in the domestication of *Malus domestica*, except possibly in the last few centuries since the arrival of *Malus domestica* in North America.

If you are wondering what this has to do with your life, consider this apple math:

M. domestica x *M. sylvestris* = Granny Smith (Australia)

When we encounter the smooth green apple, sheathed in plastic at the supermarket, Mia tells me that she likes the tart Granny Smith. I consider my tree climber with wonder. She is no doubt following her father's tracks; when Michael was a graduate student, he was a Granny devotee. The hardy apple originated in 1868 from "discarded apples" that Mrs. Thomas Smith of New South Wales, Australia, brought home from Tasmania.

Interestingly, the chloroplast genome of the Granny Smith's closest genetic cousins is from *Malus sieversii*. The supermarket Gala is of an even wilder hybrid origin, with approximately 28 to 40 percent of its genome derived from *Malus sieversii* and 25 to 37 percent derived from *Malus sylvestris*. A cross between the Golden Delicious and the Cox's Orange, the Gala was developed in New Zealand in 1934. The ancestral history of apples illuminates to me that what is all too familiar today can have surprisingly wild origins.

Scouring scientific literature, I gained an even more nuanced understanding of current knowledge: The closest relatives of the modern apple are the *three* crabapples—*Malus sylvestris*, *Malus baccata* (Siberian crab), and *Malus orientalis* (Caucasus region)—and the wild Kazakh apple. While *Malus sieversii* is believed to be the original progenitor, during the domestication process, the apple acquired a large number of genes predominantly from *Malus sylvestris* and these other crabs. So many genes were integrated from the European crabapple that they account for roughly a third of the apple genome. The apple is thus not a hybrid cross between *Malus sieversii* and *Malus sylvestris* but rather a descendant from *Malus sieversii*, changing as it was domesticated (either intentionally or not). When the cultivar apple came into contact with crabapples, especially the now-endangered European crabapple, its genome acquired a number of genes from these crabapples too.

As weeks dragged on and my request languished, I began to suspect that I would not get a response. Given that the Zhongar-Alatau National Park gets fewer than fifty visitors each year, and the bureaucrats are presumably not overworked, I could safely assume that my request was not received favorably. Since Kazakhstan and Russia share "security arrangements," which means that Russia could arrest an American national in Kazakhstan, I could hardly press any further (in case my request was absurdly perceived as being less than innocent). After all my preparations, however, this felt like a serious letdown. Looking instead at photographs of the Tian Shan, I found it striking how closely they resemble the Sangre de Cristos of my own hometown.

Chapter Seventeen

Love Notes from Nature

JULY

Author's garden in bloom. Photo by Priyanka Kumar.

I

One sweltering July, an old woman who lives diagonally across from us crossed the dirt road and approached me with a peach in hand. "It's from a peach tree that grew in my compost pile," she said.

I'd seen this petite woman, with snow-white hair, from a distance, but never before had she spoken to me. I gulped, getting over my surprise before gratefully accepting the small peach, which turned out to be tart. Still, I relished listening to my elusive neighbor. She told me that over the years, several peach trees have sprouted in and around her compost pile—then she grinned with an exultant air, as though slyly imparting a secret. I envied her volunteer peaches. Volunteer fruit trees are love notes scribbled by nature, and I yearned for sweetheart signals of my own.

On a July evening, nursery owner Gail Haggard took me and a few others to visit the garden of a woman named Clare Gardener (yes, that is her name). Clare grows all her salad greens in a deep oblong bordered with straw bales—essentially the structure we use for our compost pile. It may sound bizarre to think of the compost pile as a garden bed, but I call such inventive approaches "wild gardening." In a time of drying rivers and depleted reservoirs, when our farmers forego growing winter wheat or must halt their growing season in September because the city or state engineer has shut off their water supply, Clare grows greens throughout the winter—the straw bales keep her "garden bed" moist and warm. I marveled at her crisp lettuce and bountiful kale, standing tall and sheltered from the sun and cold by straw bales.

Straw-bale gardening is only one approach, but it reflects the innovative spirit we need to navigate the ravages of global warming, such as soils drying from increased evaporation even as there is less water to irrigate them. The rainfall we now get is below New Mexico's long-term, thirty-year average, but the precipitation in the state has actually stayed relatively unchanged over the last century. What has changed between 1896 and 2020, according to EPA data, is that our state is 2 degrees Fahrenheit warmer. Which means that increased evapotranspiration, sometimes called the invisible side of the hydrological cycle, is causing plants to consistently lose more water. In many parts of the country, soil and surface water sources such as reservoirs are experiencing greater evaporation, and we have reduced stream flows.

The term "wild gardening" suggests how we might reenvision growing plants in a changing climate and grow some perennial edibles for ourselves *and* for birds and other wildlife. When at least 75 percent of what we grow are native plants, we embrace biodiversity and spark ecological connections for moths and caterpillars who depend on natives, and the birds who feed these invertebrates to their chicks. My garden consists mainly of native plants—our *Fallugia paradoxa*, Apache plume, is a native member of the same rose family to whom apples belong. Sturdy and ethereal, Apache plume is at its finest when summer breathes her last and the shrub's white flowers give way to deeply pink styles with feathery plumage. In the midst of native plants, we add a few edibles in keeping with our quest to grow some food and perhaps because there's nothing like edibles to keep kids engaged in the garden. Most fruit trees are non-natives, but they foster our connection with nature while also providing bird habitat.

Birds and fruit trees are intimately connected. When birds migrate in the fall, their droppings deposit seeds in unexpected regions. Some of the birds I cherish, including bluebirds, northern flickers, and scarlet tanagers, relish both the apple and the tree it grows on. The biologist Richard M. DeGraaf and research associate Gretchin M. Witman have tabulated the names of thirty-one species of birds who use the fruit, seeds, buds, and flowers of apple trees. These include my long-standing friends: hairy and downy woodpeckers, kingbirds, orioles, cedar waxwings, tufted titmice, hummingbirds, towhees, and evening grosbeaks. DeGraaf and Witman note that the apple tree is "a preferred nest site of many species," such as the robin, great crested flycatcher, and red-eyed vireo. Recently, I came upon an exquisite red-breasted nuthatch who was letting out an insistent call to defend his nest site, drilled into the ample trunk of an old apple tree. DeGraaf and Witman add that "the fruits are eaten by many birds. Grosbeaks (evening and pine) also eat the buds, to the dismay of the orchardist."

Orchardists are naturally dismayed by any damage to the harvest, but birds are also brilliant helpers, consuming caterpillars by the dozens and reducing the need to spray against pests. "Let bluebirds work the morning shift, and get bats to take over at night," Gordon Tooley says. So, the key is to plant food for ourselves *and* the birds. When I keep two birdbaths watered through the summer, I inadvertently invite over mosquitos, but bats also follow, with one or two whizzing and arcing overhead at twilight, hopefully catching the mosquitoes, who always find me attractive. On the east side of the garden, a prolific bird cherry tree steers birds away from our edible cherry. Our bird cherry is popular in all seasons, serving as a nest site for robins in the spring and a snack for juvenile deer in the winter.

When I visit an old homestead or ranch, I keep an eye out for the orchard. By design or because of the passage of time, many of these orchards today embody the "wild gardening" spirit. One summer, we stayed overnight at the DeHaven Ranch in northeastern New Mexico. On a Sunday morning, before breakfast, I spied apple trees in the distance and began to walk toward them through thick grass while keeping an eye out for rattlesnakes—the day before, I had seen the intact skin of a prairie rattlesnake, including the head with an impression of eyes bulging out. The apple tree that drew me to the grove was at the center of a wide line of mature trees, hewing a rise in the land. Our lovely hostess, Stephanie Brock, who grew up on this ranch during the fifties and sixties, followed Michael and me to the orchard. She estimated that the apple trees are a hundred years old. In the 1950s, when she was a little girl, the trees already bore good fruit. Stephanie's cousin Les soon joined us. Gray eyes sparkling, he recalled the fruity summers when he and his cousins raced through the land, working it but also relishing the fruit. He told us that back then, the elders had a maxim: "One boy is the equal of a farm boy. Two boys are half a farm boy. And three boys are like not having a farm boy at all."

From an apple tree issued terrific scraping-buzzing sounds. Not one but five Bewick's wrens hopped on the boughs at the center of the tree, tracing a rough pentagon with their frenzied circumambulation. A bounty of greenish-red Jonathans hung from the tree; they would ripen in October. Beyond the apple tree towered a mature pear tree, and a western kingbird lorded over the orchard from atop. With a soft gray back and lemony underside, the kingbird fly-catched to his heart's content. It was early in the day, and we swatted mosquitoes on our arms while a handsome nighthawk executed aerial dives over an adjacent creek. To complete the quartet of bird species, a flicker flew over and took his place atop an apple tree in the back.

Stephanie and I walked along the line of apple trees, which hadn't been pruned in a few years, and I absorbed the character of this unruly but much-loved orchard. Soon we came upon two massive trunks, lying next to each other, collapsed on the dirt. Years back, one of these trees had grown so weak that one morning when a bear climbed it in search of breakfast, a large branch collapsed, taking the bear down with it. The next morning when the family saw the felled branch, a large pile of bear cobbler was found next to it. After those two mature apple trees died, Les put in a young tree to replace them. But this tree is struggling. The apple tree has a long period of juvenile development and growth, which may be compromised by our drought. Since 1950, the family has relied on a gravity-fed flow of water to irrigate the trees. That may have worked in the old days, but over the last couple of decades it has gotten too dry to rely on such natural, graceful ways.

A beautiful crabapple tree once grew at the end of the row of apple trees, and I saw its gorgeous upright skeleton. "Your folks knew what they were doing by planting the crabapple there," I told Stephanie. Crabapples make a bevy of blossoms and attract pollinators, who then buzz over to pollinate apple trees. Stephanie recalled that once when she was young, she'd seen a swarm of bees moving like an organism. "They sounded like nothing I'd ever heard before, and this 'organism' headed for the orchard. I raced over to tell my mother." By the time Stephanie and her mother ran back, the bees had settled en masse into the crabapple and were busy doing their ancient work.

Places such as DeHaven Ranch are anomalies now, and even here, the old fruit trees have collapsed or are singing swan songs, and the young trees that were put in somehow haven't grown robust. It is almost as though the trees are waiting for us to rethink our ways and rewild our hearts. In a recent report, the Intergovernmental Panel on Climate Change cautioned that preparing for future threats—such as dwindling freshwater supplies or irreversible ecosystem damage—will require "transformational changes" and "rethinking how people build homes, grow food, produce energy, and protect nature." I see wild gardening as a way of rethinking food—growing what we can, foraging a little (while leaving plenty for wildlife), and refusing to poison the earth. As we journey into the future, birds and bees, and gardens and compost piles, will be among our closest allies.

II

Rewilding food begins with assessing the foundation—the soil. If the soil is neglected or deadened, can it be revived to its native state as a living organism, alive with microbial activity? To read the mysteries of nature, we could bring an investigative spirit to the outdoors. Sometimes, micro-observations are the most revealing. Every year, I bring in some essential herbs to extend their growing season. One fall, I brought in six basil pots and placed them on windowsills under differing conditions. I tucked the smallest pot into another pot, taller by some five inches. To my surprise, I found that this small pot grew the lushest, greenest basil; on inspection I noticed condensation inside the "cover pot," which acted like a coat to keep the basil warm and moist—conditions this herb prefers. I wondered if similarly elegant solutions could be used outdoors and realized that, in our extremely windy springs, hardy brush such as chamisa, or rabbitbrush, could be used as windbreaks to shelter fragile plants such as young blackberries.

We haven't had much luck with blackberries: The canes die or don't bear fruit in our dry soil. Perhaps we aren't giving our blackberries enough love. One spring, I all but gave up on blackberries and instead began growing strawberries in pots, in our terrace. To my surprise, we got a beautiful harvest of bright red, heart-shaped berries, which kept coming almost until October's first frost. Love notes arrive mischievously, when they are least expected.

When I am able to touch this land's potential, and watch my children picking perennial blackberries or strawberries, my father's spirit becomes a living presence in my life. He exemplified Lao Tzu's words,

"Lay low to stay on top." I now see that he "lay low" most of his life, and only after he was gone did I realize the extent of his influence. He would never know that his interest in growing fruit would become his legacy. Some people have the good fortune of living near their past. They can visit an ancestral home when they like; their great-grandmother might still be alive. But when all you have of the past are shreds and scraps, you find yourself considering those with sustained attention.

One way to have a conversation with the past is to allow it to live through you. Pika likes to take her friends out in the garden to show them what is growing. When I hear complaints about the girl who "ate up all the blackberries," I remind my kids that it is a joy to share fruit. In the early days, Pika was quite reluctant to share our best fruit—or her favorite toys—but over the years, she has internalized generosity.

One summer afternoon, Pika and her best friend sampled all the viable peas, cherry tomatoes, and strawberries in the garden. Then she took her friend to a tree that grew meaning instead of fruit: "This is my birth tree," she told her friend, suppressing a smile, not wanting to give away that the tree delighted her. She gestured toward the golden rain tree we had planted in honor of her birth.

"What is that thing buried under the tree?" she raced over to ask me.

"Your placenta," I said.

She nodded with satisfaction. Through this ritual tree, we have marked our living connection to the soil—our love for each other is tied into our love for the soil. When Mia was a newborn, we planted a different variety of golden rain tree to honor her birth as well. We eventually put in a wooden swing nearby, a perch from where to watch the trees grow.

Together, Pika and I planted sunflower seeds all around; in the fall, the ground around "Pika's tree" springs to life with riotously yellow flowers nodding in the breeze and goldfinches hovering over dried seedheads.

This past summer I didn't use gardening gloves—two sets were worn through, riddled with holes. I felt the soil with my fingers, letting it sift through. This was therapeutic until I came across something hard and irregular—a nugget of asphalt. From time to time, we pick these disturbing chunks from our soil; some may have been neglectfully thrown here when our house was built, around 1991, or they may have rolled downhill more than a decade back when a neighbor uphill from us repeatedly had his steep driveway paved.

After feeding our soil organic homegrown compost so that the berries, flowers, and herbs can benefit from a full range of plant nutrients, it is jarring to discover that the soil is studded with asphalt. When I find discarded petrochemical nuggets in my garden, there's little to do but patiently pick them out. Pika stares at these chunks with curiosity; when she's digging a "river" to run from one part of the garden to another, she occasionally finds asphalt lumps for me to throw out. It's depressing to watch her handling these lumps. But she is unfazed and goes on steering the course of her acequias. Sensing her absorption in her network of waterways, a light turns on inside me: My dismay transmutes into a desire to heal this land.

Strangely, even eating apples is no longer a harmless pursuit. Today, while asphalt has contaminated some residential soil with heavy metals such as cadmium, most agricultural land is heedlessly sprayed with poisons. Corporations and the toxicologists they employ swear that chemicals such as glyphosate are safe for human consumption—though scientists have seen strong links, for instance, between glyphosate and non-Hodgkin lymphoma. Bloated with dollars, corporate spokespersons stay mum about the "cocktail effect" that consuming fruits and vegetables sprayed with multiple chemicals wreak on our health. I recently spoke to a career toxicologist who has testified as an expert witness and helped throw out court cases in which plaintiffs asked for compensation for the debilitating effects of glyphosate. The toxicologist spoke eloquently about how there is no definitive causal link between glyphosate and cancer in published studies. "Glyphosate is designed only to kill plants," he said.

Troubled by his smooth answers, I asked at last if he eats organic food.

"Yes, I eat organic," he said.

"Why?"

"It's better. The produce is better."

Those of us who can afford to eat organic do so, privately, which is a form of elitist resistance. But I dream of the day when we will root out systemic practices that denature our food and soil—so that all children, regardless of socioeconomic status, can eat apples that haven't been sprayed with poisons and grown in dead soil.

"Soils are gray, not black anymore," Gordon Tooley said ruefully, when we spoke about how, nationally, we are in an evaporative water cycle—too much bare ground causes water to run off instead of infiltrating the soil. If we could change land conditions so that water is increasingly absorbed by the soil, the water cycle would better nourish plant communities. If we remain stuck in an evaporative cycle, climate challenges such as extreme rains will continue to devastate the land with flash floods and other disasters. Perhaps I repeat like a refrain that this has already happened in the La Cienega area where the Real Orchard stands. Flash floods spurted soil into the historic acequia that irrigated the fruit trees on the west side of the property; when no one repaired the clogged ditch, the mature fruit trees were starved of water and they died.

III

I need my day to work, so gardening is fitted into early mornings, accompanied by the silvery tinkling of a congregation of bushtits; at lunchtime, when dozens of crows caw blackly in the cyan sky; and after an evening walk, with a pair of orange-and-black spotted towhees scratching dirt nearby. At twilight, with the sky smeared mauve, I sometimes shovel compost writhing with worms and land-dwelling crustaceans out of our pile, transfer it to a rusty wheelbarrow, and spread it around the plants and trees that need more love. Gary Snyder observes that "not only plum blossoms and clouds, or Lecturers and Rōshis, but chisels, bent nails, wheelbarrows, and squeaky doors are all teaching the truth of the way things are." Over the years, my rusty wheelbarrow has taught me much about the way things are. It is a refuge from the computer, but it also shows me how little it takes to become a refuge. One rainy afternoon, when the wheelbarrow happened to be upside down, I turned it over to find a baby rabbit quivering in its makeshift shelter. Pika was a toddler then, just woken from her nap; I called her over, and we had a long, juicy look before the cottontail trembled with a sudden spasm and darted off.

While I am working the wheelbarrow, Michael is patiently hosing down our piñon pines, which are regularly attacked by scale insects, or mulching and watering the conifers and fruit trees scattered around. With compost and mulch as allies that help the soil retain moisture, and birds and bees doing the rest, we are growing our Alice-in-Wonderland apple among others, and three crabapple varieties—and love-note apricot trees that have sprouted from seed. This wild gardening, tucked into the margins of our lives, spurs me to appreciate failing blackberries as much as thriving strawberries and to gently embrace the way things are.

IV

I forage fruit on weekends, and the week that stretches ahead is flavored by those trips. On the mornings when we hike along shaded trails, my eyes and ears remain open to birds—flighty juncos, hairy woodpeckers who belong in a raucous band, and the heart-opening western tanager. Slanted sunbeams caress the Douglas firs as I take in the fragrance of tart apples, glossy rosehips, or muddy strawberries. When I forage, I take a little from where there is abundance—for resident birds will need these fruits and berries in the winter. There is nothing more delightful than finding a patch of wild raspberries in wet areas near streams. One summer, we were hiking in northern New Mexico when Mia found a patch of red berries along the old Tolby Creek trail in Cimarron. "Are these edible?" she asked.

I tasted a pinch; it was a love-note raspberry. *Rubus neomexicanus*.

Mia had been dragging her feet on the hike, but the raspberries all at once brightened her mood and she busied herself looking for more. A keen observer can find this shrub, in the rose family, along our hillsides, canyons, and streambeds. I can't think of anything store bought that could have spurred such transformation.

The next July, we spent a week in Taos, where it was uncomfortably hot, and the air conditioner in our car was broken. Pandemic-related supply chain shortages meant that an air conditioner part we had ordered weeks earlier hadn't arrived. We don't have an air conditioner at home (and don't intend to get one), so we fully experienced the extraordinarily sultry summer when two-thirds of the West was said to be in extreme drought. In Taos, we took refuge in the Columbine-Hondo Wilderness. One morning, we started on the fittingly named Italiano Trail, where a profusion of wildflowers were intoxicating: softly purple *Campanula rotundifolia*, common harebell; wild roses and Indian paintbrush; *Lonicera involucrate*, twinberry, with attractive but bitter fruit; the white flowery sprays and giant leaves of cow parsnip; effervescent and pinkly violet bee balm; poisonous but strangely seductive *Aconitum columbianum*, deeply purple monkshood, used in medieval times to induce the delirious state of a werewolf; and the queen, *Aquilegia caerulea*, the blue columbine, trussed up to astonish, skirting the riverbank in heavenly blue clusters.

In the midst of a drought, it was invigorating to bathe in colors so wildly different from our usual tan, olive green, and sky blue. I wanted to lose myself in the pure space of flowers and swallowtails, but there were streams to cross. We were in strange terrain for we made eight stream crossings over swiftly flowing water, hopping over slippery rocks to get to the end of the trail. Mia, who has been athletic since probably the day she was born, reveled in the stream crossings. It had rained hard the evening before and the air was palpably humid, with the path sometimes caving into slippery squishy mud. The Rio Hondo effusively sang a love song, her water cascading down gray stones; a tributary of the Red River, which is a tributary of the Rio Grande, the Rio Hondo exuded sun-sparkled glory.

The singular trail was made all the more striking when Pika, who closely observes the ground and its critters, spotted a patch of *Rubus strigosus*, wild red raspberry. The raspberries tend to grow along sunny patches near streamsides. At first their bright red comes as a surprise, then a child bends to pick one and confirms with a smile that it is a raspberry. Soon they outdo each other to pick more and generously offer us the soft rubies. The exertion of a long hike softens. With sweetly sour berries melting on my tongue, I feel as though the forest is speaking to me. We go on wandering and a look of quiet concentration comes over the children's faces. The contours of my body dissolve and

I shapeshift into the child who would lose herself in bamboo groves. The children and I peer into the greenery, not wanting to overlook a treasure—a cluster of raspberry canes lurking behind some young Gambel oaks. We are in nature's fruitery.

It is easier for small people to spot such diminutive berries. The flavor, a bright red burst on a parched summer day, is as tart as it is sweet—a luscious love note if ever there was one. Tied closely to seasonal eating, foraging allows us to experience peak flavor and nutrition in fruit. Over time, sustainable foraging has transformed how my children relate to the land—they pick clean the strawberries from our wild garden (if they can get them before the squirrel), and stretch and climb every which way, ducking spiders and snakes, to pick apples, apricots, plums, and raspberries from unharvested groves or forests. Raspberry canes conveniently grow alongside certain trails in the Santa Fe National Forest, tugging Mia and Pika to the end of the hike. The canes are short, in our parts at least, and, as Euell Gibbons neatly describes this plant, "the leaves are compound, with from three to five leaflets, irregularly notched around the edges, green on top and downy white beneath." The berries are ripe in July and August, and none are left to bake with for they are consumed right away.

At the Italiano Canyon trail, the eight stream crossings and the raspberries ushered us to a stately aspen grove where silvery-white trunks congregate like venerable elders. We rested on a thick log, listening as a flock of Steller's jays with pointed black crests scolded a poor robin, who at last took the loud hints and fluttered away. On the way back, I lingered to appreciate the love notes written by water: ferns in cracks between metallic-gray rocks encrusted with bright-orange lichen; painted ladies and swallowtails sailed past my face. Pika pointed out several small moths and butterflies, which she expertly cupped in her hands to show us before releasing them. Water, sun, and soil have lovingly poured this lush extravagance into being. Nature held me in a green embrace. I felt drunkenly smitten and was soon daydreaming about a stroll in the Italian countryside, across flower-decked meadows in Sienna. . . . I caught myself when, at the trailhead, I walked out to my bleached olive-green-and-tan reality and Smokey the Bear warning about the risk of wildfires.

Foraging has transformed me into a member of the forest community, which is as dear to me as my human community. Early autumn means that rosehips are ripening to fleshy orange and red hues; rosehips are über-rich in vitamin C, but the practice of searching for them can be equally healing. Along shady trails, Mia and Pika help me pick rosehips—though some clusters are infested with inchworms—while keeping a lookout for the last wild raspberries of the season. I pause before the chanterelle mushrooms, noting their intricate patterns. Alexandre Antonelli likens mushrooms to "apples on an apple tree, with the difference that the tree is practically invisible to us." On our foraging journey, we can see the topsoil but not the invisible layers or mycelium networks humming underneath. These networks may be hidden, but they aren't small. In fact, the world's largest organism is believed to be a honey mushroom in the genus *Armillaria*; the fungal threads belonging to a single individual in Oregon can weigh as much as thirty-five thousand tons!

Magnetized to this wilderness, we returned in the months that followed. One afternoon, after a hike, we were driving past the Italiano trail when we chanced upon a flesh-and-blood bear, a juvenile attempting to cross the road. An SUV was speeding toward the young bear and, as it came dangerously close, Michael honked to alert the driver. The SUV slowed at the last second and avoided hitting the bear, who hurriedly slipped into the other side of the forest. Some busyness must have kept the driver from noticing a handsome, live Smokey right before his eyes. Nature's love notes aren't only her juicy fruits but also her swarthy animals. When we slow our rhythms and explore her forests, surprising riches percolate into us and rewild our inner lives.

Chapter Eighteen

Summer Lake

AUGUST

James Foster's historic house in Summer Lake, Oregon. Photo by Priyanka Kumar.

I

The Pacific Northwest, especially the Willamette Valley of Oregon, is among roughly ten regions in the country that harbor the most apple diversity. I was researching historic orchards in the Summer Lake region of Oregon when I discovered another case of an orchard razed that depressed me. The orchard had been planted by James Foster, one of the first homesteaders in Summer Lake in the 1870s (he and his parents had sold their land in Missouri and endured a harrowing journey on the Oregon Trail in 1845). Foster went on to grow a seven-acre orchard with peach, apple, pear, apricot, cherry, plum, and prune trees. He consistently had fruit, and "it was not wormy." He also grew hazelnuts, hard- and soft-shell almonds, and walnuts. It is believed that he had "one of the finest orchards in Oregon," and every year he sent a box of his legendary peaches to the staff of the *Lake County Examiner*. People came from Northern California and elsewhere to buy Foster's fruit "by the wagonload." At the turn of the twentieth century, he was rewarded with exceptional years. In 1907, "he sold more than seven thousand pounds of peaches, several thousand pounds of prunes, and dried a large amount," as Teressa Foster writes in *Settlers in Summer Lake Valley*. He was known as "Uncle Jimmy" in the community, and he liked to corral his guests early in the morning "for tours of his barns, horse stables, and orchards."

The current owners of Foster's historic home simply cut down his magnificent trees and plunked down a prefab house in their stead. I drove over to the property one Sunday morning and saw that Foster's two-story white house was nailed shut. I tried to imagine the dance floor on the second level of the ethereal house, which had been the site of many community gatherings. Foster's wife, Elizabeth, "loved music and parties" and persuaded traveling musicians to stop for a while and play at the lively dances she threw in her ballroom.

From a window of the slate-gray prefab, a woman's face peered out. A community member had telephoned her in advance to say that I would stop by. But no one stepped out.

"Apple trees were the first things the settlers around here put in," one resident had told me, "even before they built their homes." But Foster's passion had caused his orchard to become storied; it nourished his growing family, and he took care to harvest the fruit himself and made it an annual practice to gift fruit to others as well. Foster understood the primal importance of fruit to a desert dweller. The high desert landscape with an elevation of four thousand feet can be an unforgiving place to grow trees in. So, while he indulged his many daughters and built them a springhouse for their baths, in times of drought he refused them bathing water, preferring to divert precious drops to the orchard and vegetable garden.

In the letters of his oldest daughter, Anna Foster, a hit recipe featuring dried apples—Farmers Fruit Cake—would be found. "All fruit had to be preserved by drying until Mason and Ball glass jars were introduced," Teressa Foster writes. Dried apples, prunes, and apricots were essential to vary the "monotonous and limited diets" of the time. I felt disheartened to see that not a single one of James Foster's beautiful fruit trees was left intact. A ten-year-old boy wandered out of the dreary prefab. The land was exposed and there was little shade. A tractor stood next to a barn, where hay was gathered in large bundles to be sold to a corporation that will ship them to Dubai or Japan. There was enough land that the orchard needn't have been razed to make room for the prefab. Then the boy might also have experienced the rasa of weaving through fruit trees. The French philosopher René Daumal sees rasa as *savor*—"if we recall the moments of intense aesthetic emotion that we have experienced, a certain 'savor' will come to mind: and we will see how and why this gustative image asserts itself." Daumal goes on to describe savor as an immediate experience that leads to a cognition "shining with its own evidence" before circling back to an ancient Indian understanding of the concept as "conscious joy."

Unable to experience any rasa at the site, I decided to leave and walked across the street. Down below, by the stream, I noticed a volunteer apple tree growing; it was bountiful but too far down for me to access. Later, in a neighboring property, I got to taste apples from trees that had been planted by Foster's descendants—Ralph and Princess

Foster. But the fruit was sour or bland, perhaps because the subsequent owners didn't know how to care for the trees.

During the Oregon trip, while inspecting a dead snake at the boundary of the Turner Orchard, a rare surviving historic orchard in Summer Lake—once called "the garden spot of southeastern Oregon"—I was perplexed when the drone of an ATV procession careened by me. They were displaying American flags and a sinister black flag with a skull filled in with American stars and stripes—the punisher flag, a popular right-wing symbol with a likely nod to January 6. The ATV drivers may have been on a joy ride while also patrolling the area to check out outsiders like myself. I couldn't help but think that these drivers are distracted by fabricated threats while the real threat looms ever larger. Wildfire smoke blotched the horizon and hung in the air during most of the month that I spent in Oregon; as the air quality worsened, I shut the windows of the cabin where I was staying, canceled field trips, borrowed an air filter, and turned it to its highest setting.

Understanding fruit trees, and how arboreal history is intertwined with our own, is a way to appreciate not only the critical role that mature trees play in mitigating climate change but also why genetic diversity is essential in this fight. With fewer tree species—a characteristic of our era of habitat loss and biodiversity loss—orchards and forests have grown less stable and more vulnerable to pathogens. Orchards need more varieties of apples to become resilient to climate change, and biodiverse and old-growth forests could better resist climate challenges, including severe wildfires, warming-related beetle infestations, and other debilitating diseases trees now face.

II

Paisley, Oregon, is one of those sleepy towns (population: 250) you pass through on a road trip without as much as pausing. A resident I knew had told me about two birding spots by the Chewaucan River and Lover's Lane. So, we did stop. And saw a warbling vireo, then a juvenile line snake busily crossing a dirt path, and raptors wheeling above us; we were a bit off with the time of day (mid-morning) and season (the height of summer). The weekend before, I had attended the Mosquito Festival parade in Paisley and met the friendly librarian Jan Murphy. "Come over to the library someday," she'd said. "We have good Wi-Fi."

After a poor morning of birding, I thought it might be nice to take my girls to the Paisley Library so that they could read inside in case we weren't allowed to borrow a book. Jen warmly ushered them into one of two kids' rooms. The library was a narrow blue-green structure, the size and aspect ratio of a large trailer. Half of a lush, prolific apple tree caressed the roof, and this tree was on my mind. A Paisley resident had also told me that an old Gravenstein apple tree grows on her property. The Gravenstein is an exquisite Italian apple from the early 1600s, and it is believed that Russian settlers brought this apple to California in the 1820s. While the kids were browsing, I asked Jan if she had any material on the history of heirloom fruit trees or orchards in the area. She shook her head. "Not much has been written about this area," she said. "Though I wish someone had written about the heirloom fruit trees."

Her supervisor, Marsha, concurred with the assessment. She happened to be here, though she is the librarian at the Lakeview Library, an hour away. Marsha had been quiet so far but now she began to talk. "My great-grandfather Harry planted an orchard," she began. A pale woman with shoulder-length reddish hair, perhaps in her mid-fifties, she wore a checkered skirt suit. "When we were little, he would go out and get apples for us. Even when he was very old, he would go out and get an apple for us. He was in his early nineties when he died—in 1983." She paused thoughtfully. "After him, the trees weren't taken care of, and there's very few left now."

"That's what happened to the trees Marty's granddad planted," Jan said, referring to her husband, Marty. Then I listened to Jan's story about the McIntosh apple seeds Marty's granddad had brought in 1914 from County Cork, Ireland, in his sole canvas bag. A sheep herder, he also put in an orchard in his new ranch in Oregon, before World War I began.

When Jan and Marty married in 1970, there were six to eight apple trees left.

"I didn't realize the significance of it," Jan said. "We just picked the apples." She paused for a long stretch, staring at the library counter.

"Ten years back we realized the significance of what we had lost," she said at last. All the apple trees are now dead. Jan's sister has a plant nursery and recently said to her: "I can't believe, Jan, you didn't get grafts from the old apple trees."

As I stood there, struck by the resonances with my thinking about apples, Marsha began to move about briskly, as though she were in the process of deciding something. She had been on a visit to the local library, where she was a supervisor, but now she began to collect her things, and her keys.

"Maybe I'll take grafts," Marsha said suddenly, causing me to look up. "I'm going to go there right now," she added.

When I looked at her, puzzled, she added, "To the orchard. I'll check up on the trees."

"Should I wait here for you?"

"You can come along."

I gathered my troupe, and off we went, following her gray car. She stopped just two houses from the main drag, where I had stood to watch the Mosquito Parade on Saturday morning. There is only one brick house in town—and it belonged to Marsha's grandfather. Now it belongs to her cousin, who has rented it. As we walked into the yard, we were greeted by fruitwood from a dead apple tree; it lay on the bleached grass.

"This was alive two years back," Marsha said.

Close by, a mature black walnut with a dark trunk stood tall. We walked over to the side of the house and came upon an apple tree that looked half-dead though it had three prolific branches. "This one is still alive!" Marsha cried out. Then she quieted and her eyes grew moist. "It makes me want to cry."

She held the tree almost in an embrace, so glad was she that it was still alive. Her father had passed on this past February, she later said. "If he had lived a little longer, I would have asked him about these apple trees.... He would have known."

There were three other apple trees remaining and one large apricot, which had leafed out fully and had a beautiful structure but wasn't bearing fruit.

"The trees once stretched all the way down to the bridge," she said, indicating the entire block and more.

"Wow," said Michael. Earlier in the morning we had driven down to the bridge, looking for access to the Chewaucan River.

"So, there were at least a hundred trees?" I clarified.

Marsha nodded. She hadn't been back here in some time, it seemed, and my asking about heirloom trees had opened some yearning within her. "I am going to look up some family history and do some research . . . and tell you more," she said. When Marsha was seven, she would hear the elders talk and those conversations fascinated her. She wished she could recall the things she had heard. "With my father's generation, people moved to the city," she said. Her father moved to Portland in the early sixties and, later, Marsha and her sister were born there.

"So, that connection to the land was lost?" I asked.

She nodded. "Exactly."

Chapter Nineteen

The Hidden Rose

AUGUST

Apple orchard in Kulu Valley Pass, Beas River, Himachal Pradesh, 1983. Photo by Mary Binney Wheeler. Mary Binney Wheeler Image Collection, University of Pennsylvania.

I

As August drew to a close, some of my favorite apples began to ripen. I acquired bucketfuls of Yellow Transparent and July varieties, which can be lip-smackingly tart but are good to cook with. Under the blazing sun, I would find myself chatting with an apple farmer about how his Wolf River apple tree from 1910 is doing ("It has maybe ten years left."), and how his apricot trees from the 1860s were still going strong. It heals my spirit—it's like getting a dose from my Himalayan days—to weave apple lovers into my community.

Apples have long struck a chord, and the apple symbolism in our past extends across cultures in Greek myths, early Christianity, and even in Buddhist traditions. In ancient Greek culture, when a man threw an apple at a woman, it was seen as a proposal of marriage. If she caught the apple, it meant she accepted. Knowing this, I avoid throwing apples at people, or catching them for that matter. The matrimonial theme resurfaces: Sappho compares a bride to a ripe apple blushing on a tree, just out of a picker's reach.

Mother Earth is said to have given her grandson Zeus and his bride Hera an apple tree that would bear the golden apples of immortality. Hera treasured the tree and planted it in the garden of Hesperides, her secret garden far to the west: "She put a hundred-headed dragon under the tree to guard the apples and ordered the three Nymphs of Hesperides to water and care for the tree." The past speaks to the present in surprising ways. I find myself thinking that today, those nymphs might struggle to satisfy thirsty apple trees. Like third-generation orchardist John Trujillo, Hera might fear that it's growing too hot to plant more fruit trees. "It's just too hard," Trujillo had said, feeling discouraged by drought and global warming. It hurt to hear that Trujillo plans to stop planting fruit trees, not least because he grows the most sensuous pears with caramel notes. "We had a small harvest this year," another apple farmer, Taylor, told me. In mid-September, her farm has already run out of apples. Taylor and her husband have five hundred apple trees, and two kids. She wondered if the frosts and hard rains this spring knocked off the apple blossoms.

Dante saw apple trees as life giving and likened their absence to a hellish landscape. "Not apple-trees were there, but thorns with poison," he writes in *Inferno*. Dante's imagery is ecologically prescient. Since the beginning of the industrial age, we have progressively scraped away at our landscapes until many could be compared to small and large infernos. We've now reached a point when Mother Earth totters, and climate fluctuations haunt our lives.

It is a task not only for orchardists and farmers but also for the rest of us to cultivate a climate ethic. This could mean a practice of nourishing the soil and growing carbon-sequestering, shade-giving trees. Imagine a community-making practice to ease the earth's burdens (like the baobabs the Little Prince roots out to save his planet): industrial agriculture, wanton development, and high-carbon lifestyles.

Then we might be able to inspire John Trujillo's granddaughters or Taylor's kids to want to go on growing fruit trees. It touches me that even in these challenging times, John goes on being generous. Today, he didn't accept full payment for the purchase of a box of Jonathan and McIntosh apples. "You like my apples," he told Michael and me. "That makes me happy." The real gift of being human is that we are compassionate and loving beings—these qualities feed our creativity and capacity to navigate challenges. The experience we'd accrue while tending to our trees would hone our climate ethic—and clarify our attempts to read the lexicon of the biotic communities we live in.

Reading the earth involves a joyous sitting in her folds and deciphering her as best we can. Today, when do we ever sit (and work) in nature, hour after green hour? There were times when we may have come close to discovering a golden mean between the ways of our gatherer-hunter ancestors and the deskbound modern life. In the eighteenth and nineteenth centuries, American orchards matured into forestlike idylls that picnickers found irresistible. Americans who worked and frolicked hard in orchards might well have drawn inspiration from how Eros himself pursued happiness. The vivid scene the Greek philosopher Philostratus draws of Eros harvesting apples offers some cues: "Some (cupids) hang their quivers on the branches and fly up to the fruit, others enjoy eating them, two kiss an apple which they throw to each other, an indication that they are falling in love, another is pelted with apples by spectators for biting the ear of his opponent in a wrestling match, others again chase the hare that eats the fallen apples, and finally some gather around Aphrodite with the first fruits of their harvesting."

When I forage with my children, I sink into the delights of radiant mornings: the splendors of trees singing with fruit; girls urgently filling cloth bags to bursting; tasting fruit until our bodies are made of apples. The magic of ancient trees wraps me in a trance. We are near the city, but this feels miles away from the hard edges of material life.

On foraging days, two tottering apple bowls sing from our kitchen counter and more fruit creaks the refrigerator. After dinner, Mia brings over her choice of apples for me to slice. First, we identify them, and then the kids taste each variety slice by slice. One week, the apple varieties got accidently mixed up, and every evening Mia tried to guess the name of the variety she was holding. "I think this is a Hidden Rose," she said. The rest of the family were skeptical, and we gasped when a knife revealed pink flesh inside. The taste is simultaneously sweet and tart; the apple sings in your mouth. It feels meditative to end autumn evenings with the girls talking apples. Discovered in western Oregon, the Hidden Rose's dark pink blossoms also give it a reputation as an ornamental tree that makes "green-gold cone shaped fruits." After watching many a child's face light up when I cut the Hidden Rose and reveal the surprisingly pink flesh, I have come to think of this apple as "hidden happiness."

The Russian Giant Crab, a buoyant salad apple, also makes a splash with its red flesh, and Pika's friend often asks for it, hoping that it will magically reappear. The skin of this oversized apple is partly bright with yellow-red hues and partly blood-red like a pomegranate. Pika's friend says the inside is like a tie-dye shirt. True. Red flesh is brushed through with pink. The tree is said to be vigorous and upright, and quite productive—and may be resistant to apple scab. The fruit tastes tart and alive, as though Rasputin himself were dancing before you.

We save some varieties for special occasions, like the Macouns we finally cut open during a solar eclipse. I find the Macoun to be surprisingly light and ethereal. Which makes sense for an apple that "drops readily and bruises easily." A cross between the McIntosh and the Jersey Black, launched by the Geneva Station in 1923, the Macoun can sport stripes and deep red coloring. As the kids raced into the kitchen with their eclipse glasses, I inhaled the aroma of the Macoun's white flesh. Michael thought the apple had complex flavors and tasted caramelized, with a hint of butterscotch.

I nodded. "It's better than dessert."

"Dessert is better!" Pika countered. "This tastes like apples."

I couldn't get the caramel notes out of my mind as I donned eclipse glasses and viewed the ring of fire. The pairing of Macouns with the eclipse felt cosmically sublime. In a sense, our family was creating our own apple mythology.

After Pika had looked at the eclipse through special glasses, she said, "It's not red."

"No, it's caramel," I said. "Like how the apple tastes."

II

As I went deeper into the universe of apples, a serendipitous experience helped me to parse out the meaning of my closeness with fruit trees, fostered in my Himalayan childhood. An old childhood friend tracked me down, after we had lost touch for decades. A. and his family were our neighbors in the state of Assam when I was between the ages of seven and nine. We both loved racing up hillsides and bounding from one fruit tree to another. A.'s father had recently passed away, and he wanted to talk; he went to extraordinary lengths to find out where I now lived. It was an electric feeling to hear from A.—one of my closest friends. The feeling intensified when the conversation with him and, later, his mother, turned organically to fruit. A.'s mother was visiting him in Virginia, and during a phone call between the three of us, A. reminded me of the thickly laden peach tree right outside my front door in our town of Haflong, the so-called land of the blue hills.

"The branches would be falling down," A. said.

"There was so much fruit," said his mother, recalling the profusion of peaches, black plums, guavas, jackfruits, bananas, and green mangoes, "that we never cut a fruit—we always ate it whole from a tree or a crate. Like the crate of yellow apples your father brought back from Himachal Pradesh."

I nodded as though in a dream, recalling the autumn when my father had taken me to visit the jewellike orchard in the Himalayan foothills. Embraced by the fragrance of the formative memory, I saw that it had been calling me my entire life. It was the seed that had sprouted into my desire to cherish apple trees, especially the diverse varieties that have persisted for centuries. My father had enjoyed sharing fruit, but I was surprised to hear that he'd had a crate of yellow apples shipped a four-day train ride away to our neighbors in Assam. The gesture reveals a deep strain of rasa in a man who was otherwise modest and taciturn. Perhaps it also suggests why, decades after we were neighbors, A. sought me out to talk about our fathers and the old days. In tears, I realized that picking and sharing fruit was a way of being, a container that held the secret sauce of those days when we lived not so much as neighbors, but as one big, juicy family.

After a few more conversations with A., who now masterfully tends his fruit-and-vegetable garden in Virginia, memories uncoiled and I recalled the varieties of fruits, especially the jackfruits, I would eat straight from trees in my Assam days. It buoyed me to fully inhabit the mysterious ways in which apples can connect us to our past, whether the past is evoked by our parents, Greek myths, Spanish explorers, Johnny Appleseed, Archbishop Lamy, or the Buddha himself. I felt deeply moved that A. had tracked me down, unfazed by the fact that we'd lost contact for twenty-five years. I had traveled long and far in the land of apples and at last felt ready to approach the childhood memory that had left an indelible imprint.

III

As a boy, the Buddha had his first meaningful meditation experience under *Syzygium jambos*, a rose-apple tree that is technically not an apple. The classic text *Lalitavistara* notes: "As Siddhartha meditates under the rose-apple tree, the power of his fierce concentration is such that it generates an energetic field in the sky." I wondered if the rose-apple tree has a scent, and if the Buddha's concentration blended into the scent cloud of the tree. Unbeknownst to the Buddha, the tree sprayed him with aerosols and boosted his overall sense of well-being.

The energetic field created by the Buddha's fierce concentration is said to have disrupted the flight of five airborne rishis who were "making their way to the Himalayas."

The goddess of the rose-apple tree overheard the complaints of the grumpy rishis, who were as frustrated as any stalled travelers, and she told them why they had abruptly stopped moving: "It is the Sakya Prince descendant of the best of kings, who shines like the dawn. . . . His power, gained from merit in millions of lives, is thwarting your [flight]." I do not know whether the grumpy rishis were soothed, or when they were able to resume their journey, but the story makes me smile. I find it a welcome relief when the frenetic flight of my twenty-first-century life is thwarted, and I fall under the spell of apple trees.

In his early thirties, the Buddha began to emaciate himself in order to gain spiritual mastery. He grew skeletal to no purpose, however, and eventually questioned his methods. During the questioning, the simple moment of pure contentment under the rose-apple tree returned to him. He recalled: "I thought of a time when my Shakyan father was working and I was sitting in the cool shade of a rose-apple tree: quite secluded from sensual desires, secluded from unwholesome things, I had entered upon an abode in the first meditation, which is accompanied by thinking and exploring, with happiness and pleasure born of seclusion."

Childhood memories may be evanescent like mist, but the Buddha's experience under the rose-apple tree would blossom into a key insight—that the quiet happiness he had experienced under the rose-apple tree was worth examining. In a humbler sense, this is what I've been doing—examining the quiet happiness I experienced as a child in a Himalayan orchard. At some point, I began to wonder why I had become so removed from that experience and how I might circle back to it. The meaning behind the memory called out to the Buddha, through the passage of years and, remarkably, set him on the path to attain enlightenment under a fig tree.

One recent autumn, the practice of walking in orchards let in a chink of light, and my elusive memory grew fluid, allowing me to walk through.

That fall morning, while picking tart feral apples in Santa Fe, my formative memory returned with the clarity of a crystal:

I was five and wandered mesmerized through a grove of mature, wide-canopied trees in the foothills of the Himalayas; rosy and gilded apples hung around me like glowing ornaments. I walked and walked until my stubby little legs ached. My father spoke to the orchardist and they went on walking without noticing me. A moment came when the walking and aching became part of me, and I no longer noticed them. I might have been walking in the orchard forever among trees who were my ancestors. Then, abruptly, we came to a house, and the orchardist led us up some stairs to the very top, where we entered a turret-like attic with glass walls.

There was just one armchair, upholstered in a checkered mustard cloth, and the man with a round, grizzled face offered my father a seat. Once he was seated, a subtle smile played on my father's face. The light in his eyes (a dream of quiet happiness?) entered my heart. In the sun-bathed room, I settled on his lap and took in a bird's-eye view of the apple trees. The owner approached me and offered a *saiv*—the Hindi word for apple. It was deeply yellow and sweet, with tones of custard and honey. I was surprised by the ambrosial nature of the apple but soon sank into taking another bite. I slung an arm around my father's neck and munched the apple, feeling like a bird gazing at the glittering, sunlit orchard spread before me. Nestled in my father's silent love, and eating the custard apple, I experienced an awakening: I saw that the earth herself had opened her great green arms and lumbered over to give me an apple-scented kiss.

My father and I never spoke about the orchard visit. I suppose that wasn't his way. We both loved fruit, and he sometimes teased me about it. Our lives were rich with gestures that were a nod to this love; during harvest times, I hovered about him worse than a housefly. Later, when we lived in a city, he usually biked by the local farmers' market or traveled by jeep to a farm to acquire rasa-filled fruits. Some evenings, he brought home an offering miraculously balanced on his bicycle. We delighted in a crate of mangoes or a whole box of lychees. Or a surprise gift of guavas that were pink inside, like the Hidden Rose apple—a joyous throwback to the fruit trees we had once treasured.

Chapter Twenty

The Songs of Ancient Trees

SEPTEMBER

Mother apple tree in the Manzano Mountains.

"Are those apple trees?" I asked. In the near distance at the Quarai ruins was a row of four or five trees. Going by their structure, shape, and height, I intuitively wondered if they were apples.

"That would be exciting," Michael said.

For two weeks, I had been hearing the word "None." As in, "There are none left" or "I don't know of any." I'd spoken to half a dozen rangers at the Salinas Pueblo Missions National Monument and the Manzano Mountains State Park and, despite their shrugs, over Labor Day weekend my family and I drove a couple hours south to Mountainair to see for ourselves. In my heart, I couldn't believe that no historic apple trees were left in the ruins that make up the Salinas Monument in the foothills of the Manzano Mountains. Considered one of the earliest sites where the apple was grown in America, the Manzano Mountains and the village at their base are named after the Spanish word for apple tree—*manzano*. A place name doesn't generally appear out of a vacuum, and a place named after a landscape feature suggests that it was meaningful in the culture, history, or daily experience of those who once lived there. Michael and I had pored over a bevy of old maps and determined that a 1794 map by French cartographer Jean-Baptiste D'Anville seemed to have been the first to refer to the Sierra Moreno as Mansos Mountains. At the time, the mountains may have been named for the Manso Indigenous peoples. The town of Manzano, dating from the first decade of the 1800s, was named for apples, suggesting that mature apple trees already grew there. An 1867 map by the American cartographer Samuel Mitchell refers to the adjacent mountains as Manzanas. A map prepared in the same year by the US Topographical Bureau officially christens the range as the Manzana Mountains. A paper published in *New Mexico Geology* in 2000 notes that "the few remaining trees" in this area "are probably the oldest apple trees in the United States"; their history stretches back to early Spanish explorers and the native Tompiro and Tiwa peoples who planted the trees for seventeenth-century Franciscan friars.

At the site of the Quarai ruins, Michael briskly returned to the car to get our binoculars and then we took turns peering at the line of trees in the distance—at the top of one tree hung a handful of fruit that looked to be apples. I felt a sting of excitement; to be sure, we would need to get closer. Our kids were climbing a cottonwood and asked if they might watch from their sturdy perch while Michael and I hiked to the fruit trees. I may not have considered the request, but intoxicated by the zing of discovery so close at hand—the trees looked to be five minutes away—I agreed.

"We'll be back soon—stay right here," I said, hopping onto a trail smothered on either side with vegetation. But the trail didn't get us closer to the fruit trees. They might well have been a chimera. Still, in my mind's eye, I saw enticing apple-like fruit hanging from high branches. Searching harder, we got off the trail and promptly got turned around in the thick vegetation. On an instinct, Michael turned to our left and, a few hundred yards later, we stumbled upon the line of four apple trees.

The fruit-bearing tree was closest to me, and a wood peewee was perched on a high dead branch. A slender charcoal-gray bird, he appeared to be resting but flew up a couple of times to fly-catch. A yellow warbler flitted about the branches from which hung yellow apples with a red blush. Michael saw an orange-crowned warbler fly away. A delicious feeling of rasa bubbled up and mingled with the bright apples and flighty birds.

"I can smell the apples," Michael said. There might have been some windblown apples on the ground, but in the thick vegetation, I couldn't see any. Earlier in the morning, we had bought two buckets of raspberries from Farmer John, who then regaled us with stories of rattlesnakes and kingsnakes in the area. Recalling the farmer's recent encounter with a coiled rattlesnake by his greenhouse, I didn't care to grope about in the undergrowth to look for apples. Orange globemallow flowers and prickly pear cacti stood out against the weedy undergrowth, mostly kochia (burning bush), and blue grama grass. Deer, elk, and bear clearly knew about these trees since no windblown apples lay visibly on the ground. We examined the towering trees, fueled by questions. How old were they? Was this quartet planted intentionally or were they volunteers? The Franciscans had planted fruit trees in the Mission Gardens—a short hike away, behind the ruins of the Quarai

Church—and hired Indigenous peoples to tend to them. Unfortunately, those gardens were nothing but a grassy mound now.

Indigenous peoples lived nomadic and, later, hunting lives in the Estancia Basin from twelve thousand to eight thousand years before present. They settled into villages or pueblos around 600 CE, growing the graceful sisters—corn, beans, and squash—and thriving to the extent that the Estancia Basin became one of the most populous areas in the region from 900 to 1500 CE. Since at least the 1400s, they engaged in dryland farming and diligently kept stores of corn to tide them through the vagaries of this way of farming. The Spanish arrived, however, with silver and gold on their minds and forced the natives to neglect traditional agriculture in favor of extracting salt from the Estancia Basin—to be shipped to Mexico to process silver.

The Franciscan brothers came to the region on the heels of the sixteenth- and seventeenth-century Spanish conquistadores, eager to "harvest" or save souls in the Estancia Basin. The Franciscans have a reputation as a mendicant order of the Catholic Church, but each friar was supplied with rich and ample food, including four pounds of fruit conserves, to see them through the six-to-eight-month journey to New Mexico. The supplies were meant to carry over into the stay at the new *convento*, and the friars were instructed to become self-sufficient thereafter. In this spirit, they employed Puebloans to grow their vegetables and orchards. Records show that the friars had a taste for fine food and, on feast days, chocolate drinks (a luxury) flowed as freely as we serve coffee today. I had learned during a ranger talk that the *convento* was built painstakingly by Indigenous women who took their instructions from Spanish engineers, and these intrepid women builders became all but fluent in Spanish by the time they were done four years later. The main walls of the church still look superb. The summer when I was at Quarai, however, hard rains had beat down against the walls, smashing some in part, and causing the *convento* to be closed off until repairs could be undertaken.

The conquistadores meant to make fortunes in New Mexico, but the abrasive boldness of the Spanish soldier-settlers irked not only the Indigenous peoples but also the friars. At the visitor center of the Salinas National Monument, I had read the observations of Fray Estevan de Perea, who came to the area in 1629 as a custodian of thirty priests, and he penned down his grievances thus: "The [soldier-settlers] take up farms not only in the cultivated fields of the natives, but even in the courtyards of our convents. . . . Into this pueblo of Cuarac [Quarai] there came a soldier with a great number of cattle of all kinds, to a place where these three neighboring pueblos have all their cotton fields, and he build his corrals . . . right on the fields . . . absolutely ruining them."

Moving about in the grass, I wondered if I was standing at the site of one of the old fields. Looking up, I realized that the apple tree next to the fruit bearer seemed to be dead. The Puebloans had abandoned the Quarai site in the late 1600s, so these ancient trees could have been volunteers, planted by bears who had gorged their fill of apples in the abandoned Mission Gardens. In the Estancia region, the real gold of the Indigenous peoples—their corn stores—would dwindle under the extractive *encomienda* taxation of the Spaniards. In the mid- to late 1600s, a drought hit—for twenty years—and there wasn't enough corn to see the Indigenous peoples through. Smallpox and intertribal war exacerbated the crisis, and the desperate Puebloans fled the region, including the church they had built and the apple trees planted in Quarai.

Miraculously, these four trees were still standing, a living testament to the fraught relationships between the Franciscans, Spanish, and Puebloans—and to the labor of the ancestral Puebloans whose contributions are too often unsung. I drew a long breath, eyeing the tree flaunting the yellow-red apples, too high to reach. In the thick of contemplating the origin of the trees, an inner alarm went off; the kids were waiting, and we must get back.

Trying to make up for lost time, we found ourselves wading in waist-deep undergrowth. I soon saw the cottonwoods where we'd left the kids—but they weren't in sight—and we still had to navigate through masses of bindweed, kochia, prickly pear cacti, and sage, whose scent wafted up on the late-summer day. The woods rose all around, prickled my sneakers, and kept me alert to the likely presence of snakes. If we'd had more time, we would have retraced our steps back to the trail. But I didn't want the girls to worry; the apple trees had "caught" me, and we had lingered longer than we'd meant to. To reassure myself, I recalled that there were only a handful of other visitors at the national monument, all elderly, and envisioned the girls chatting in a crook of the cottonwood. Still. Wading through the surprisingly thick vegetation, we at last hit upon a path that big mammals—elk, bear, and bobcat—take, for we saw their scat and the vegetation trampled in a neat line. I skirted around one pile of scat, and several more, until I hit upon a final snaggy thicket of kochia. Tearing our way past, we at last emerged into the cottonwood grove—where the girls stood at the base of a tree, sulkily asking what had taken us so long.

Earlier in the morning, Pika had been walking around the Quarai ruins and found a dead hummingbird. She placed him on a bed of cottonwood leaves and walked him over to where we stood talking with a ranger. She didn't exactly show us the bird but walked past us so that we might see his iridescent-green magnificence. Struck by the sight of a child solemnly carrying a gorgeous dead bird, the ranger mumbled platitudes about "the cycle of life."

Trying to console Pika, he pointed to a rock ledge, "If you put it over there, a snake will come get it."

Pika didn't respond—she looked solemn but serene. The otherwise staid ranger seemed to grow troubled as he told us that sometimes a hummingbird can be aggressive enough to kill a different species of hummingbird. This one had a long bill, delicately curved at the tip, and Pika later told me that one of the eyes was missing. It turned out that Pika didn't need to be consoled. Rather, she wanted to console the handsome bird over his death.

The ranger muttered, "It might have a disease, don't touch it. Or it could have mites."

Pika knew to take these precautions, and she vanished out of sight. I sensed that she was around the corner, in the shade, considering how to proceed. The ranger had planned to walk us to the site of the Quarai Church and give us an overview of the area's history. But Pika and Mia had already raced over to the church ruins and back, and they asked for permission to stay in the shade rather than follow us into the blistering sunlight.

When we returned from the church ruins, we learned that Pika had given the hummingbird a royal burial under the shade of a towering cottonwood. She and Mia had dug a hole, lined it with cottonwood leaves, and added a generous layer of cotton harvested from the tree. The hummingbird was placed on this fluffy bed and covered luxuriously with a blanket of cotton and the loveliest flowers before the hole was covered with dirt and stamped on—so that an animal couldn't easily dig it up. A spirit messenger had passed on to another realm, and the girls had honored him on the journey out. Pika was moved by the ceremony and recalled the hummingbird with affection even after we returned home a couple of days later.

A strain of affection had also coursed through me when I discovered the line of four apple trees near the Quarai ruins and the gorgeous warblers and flycatchers clustered around their branches. It was a heady experience to find trees that were presumed dead—trees that had an uncanny ability to bring history to life and were havens for birdlife. One of the four Quarai apple trees was dying or dead. Who will honor the trees when they are gone? The rangers were well-meaning but seemed to be relatively new—I spoke to five or six, and no one knew about the existence of the apple trees. If we don't know the trees, much less their significance, how can we honor these nurturing mothers, living witnesses to history, keepers of juicy secrets? Pika sent off the hummingbird into the spirit world with vibrant flowers and the song of her sensitivity. Which songs will we sing when the ancient trees leave us?

After the incandescent morning at Quarai, we headed to the Abó ruins. We picnicked under a portal, where Mia climbed a ledge to get to a pear tree laden with a couple hundred pears. We ate the crisp pears for dessert before heading out into the fiery afternoon. I was looking for a very old apple tree that Gordon Tooley had taken twigs from more than ten years back. After a walk to the arroyo where I expected to see the tree, I instead encountered a monarch butterfly and a tarantula hawk wasp. I'd spoken to Gordon on the phone before the trip and he'd told me that he used scions from the last surviving tree—located south of the ruins, at an old homestead site—to graft young, viable apple trees. Gordon recalled that he went to see the tree "a couple of government shutdowns ago" with a National Park Service archaeologist who "watched me like a hawk." He found the lone apple tree to be in pretty rough shape.

"I took twigs off that, and grafted them," Gordon said. "That's when a government shutdown happened. When I tried to give the grafted trees back to the ranger, he said, 'Forget it. We'll end up killing them.'" So, Gordon gave the five or six young trees to a farmer's family in Mora. "They're descendants of the [Abó] site."

Gordon may be a wry, salty-mouthed orchardist, but it felt like an appropriate gesture of healing when he gifted the young grafts to a farmer whose family has deep roots—Indigenous and Spanish ancestry—in the area. The farmer has since moved on, however, perhaps to Hawaii, and Gordon has no idea about the fate of the Abó grafts.

As Michael and I scoured the Abó site for its last surviving apple tree, the sun pierced our faces, and the humidity was overbearing. Filled up with the Quarai sightings, I wasn't as disappointed as I might otherwise have been after hiking and searching without sighting the elusive tree.

Perhaps it was dead. The new ranger at Abó, a volunteer who'd served here for two years, hadn't heard of the tree. She called the Quarai ranger, thinking that he might know, but he instead related to her the story of the dead hummingbird Pika had found. After the Abó ranger put down the phone, she said to me, "Stay on the trail." I nodded to assure her, wondering if her colleague had seen me in waist-high undergrowth in Quarai and tipped her off. I make it a practice to stay on the trail, and did so in Abó. I offer no excuse for my anomalous behavior at Quarai except to say that it was fueled by the intoxication of discovery.

When a beloved historic apple tree in the Pacific Northwest died in 2020, there was considerable brouhaha. Rightly so. In contrast, even the Salinas rangers I spoke to didn't know about a comparable treasure right under their noses. Perhaps with a turnover in positions, institutional memory has faded. Three gorgeous ruins remain of the Salinas settlements, and the New Mexico Bureau of Geology paper confirmed that the surviving stand of apple trees is located in the ruin of Quarai. The paper characterized Quarai as nestled in a juniper forest, ideal for apple growing since the land has natural springs. Tree-ring growth data estimated that the apple trees are from roughly 1800. Local legend has it that the trees were planted even earlier by the friars and early Spanish ranchers. If the Quarai trees are 224 years old, as per tree-ring estimates, they have outlived the Pacific Northwest's oldest apple tree, believed to have germinated in 1826 or 1827, and died in 2020, at the age of 194.

Considering the forgotten Quarai apple trees, I recalled a haunting question: Whose history is amplified and whose is all but forgotten? To widen our apple history, we may have to stretch our hearts and minds. Stretching isn't always easy, but the process could in turn inform how we envision our ecological future. It feels wondrous to know that remnants of our rich apple culture are alive with us even today. The Quarai trees are like the North Star, pointing the way to a richer way of seeing. They have been pointing for more than two hundred years and, in my lifetime, will fade into oblivion.

The next morning, we headed to the village of Manzano, where we would drive up a jaw-clenchingly steep nine miles on Capilla Peak to look for migrating hawks. Before we started driving up the mountain, I spotted the mother of all wild apple trees next to a forest road that led up to a wilderness area. The tree was in a floodplain that was smothered with grass and splashed by sunflowers. Bracketed by a twenty-five-foot Gambel oak and a shrubby juniper, the mother apple towered over all the other vegetation in this part of the floodplain. She looked to be a walloping thirty-five feet tall, and her width was grandmotherly. It was impossible to embrace the trunk all the way, and the expansive canopy bore cartloads of apples. A surge of rasa flooded through my body, and I crooned: "This is the tree I've been looking for!"

The girls glanced at each other with subtle smiles; I am not known for spontaneously breaking into song. The tree cheered us all and we enthusiastically explored it. The massive trunk was a surreal sight, with branches growing in every which direction, sometimes seeming to zigzag out. The tree's bones were stunning, though the apples themselves were nothing to sing about: They were refreshing but lacked complexity. To be fair, the first day of September is not when fall apples are in their prime. A few more weeks might mellow and sweeten them. The tree's structure was strangely endearing—she could hold all of you, and you could sink into her. Our faces shone as we picked apples from the awe-inspiring mother.

That afternoon, we hiked in the Manzano Mountains to a stunning crest, where we picnicked before watching early fall migrants like the prairie falcon, a cross between a thunderbolt and a cloud, and skittish yellow-rumped warblers. Then we drove back down to the village of Manzano. At the local church, I saw a groundskeeper at work and intuitively jumped out of the car to talk about apples. Pat Romero was glad to talk apple trees, but first he singled out the wild apple tree in the forest. I nodded happily—yes, we had found it.

Pat was born in Manzano in 1968, and his father was the longtime groundskeeper for the church. Romero Sr. passed away last month, and Pat had agreed to tidy the grounds for the church's two-hundredth anniversary celebration and dance in two weeks. "Come back and visit us then," he said. His invitation grew poignant when he added that "not even a hundred people live here now . . . everyone [the old residents] died. There are only three families in their thirties." He admitted that winters are especially rough in the village, and you have to be prepared with stacks of wood.

"Do people still use wood stoves here?"

"Oh, yes!"

Soon, Pat was directing me to the oldest apple trees in town. "They're the oldest in the country—the four apple trees in Roy's yard, by the arroyo. They bear very small apples. Roy lives in a trailer. If he's there, he'll show you around."

Following the casual directions, my eyes chanced upon a modest orchard by an arroyo. Small yellow apples from one tree had fallen straight into the greenish arroyo water. Michael walked around the property, looking for the entrance. He came upon it at the same time as someone drove up from the road. The young man, maybe twenty, eyed us with surprise.

"We just spoke to Pat Romero," I explained. "And we'd love to look at the apple trees."

"I'll ask my grandma if that's okay," he said.

We waited for several minutes, and eventually an older woman in a housedress came out and welcomed us with a toothless smile. "Roy's trees are across the road," Maria told us. "But all of them died." I felt stung with disappointment, but she added that the trees in her orchard were just as old—now, they were the oldest in town. "Make yourselves at home."

The orchard's understory was lushly green. As I gazed at a mature tree, Maria added that the trees date back to the 1800s. I asked if she had any records, and she volunteered to go back inside and look for a history book. She left us to the orchard.

In a square grassy plot bordered on one side by the arroyo and on the other by grapevines grew a handful of large apple trees. The afternoon was achingly beautiful. It might have been an idyllic scene if, just beyond the orchard, weren't dumped an extended family's worth of broken cars and paraphernalia. The grandson, a handsome youth, came out with a couple of plastic bags in case we wanted to pick apples. "We have more than enough," he said.

Touched by his generosity, I tasted an apple and remarked how juicy it was.

"They're the best," he agreed. "You're welcome to go past the woodpile to the trees over there.... We have pears, some apples, and nectarines."

After he had left, Mia and I walked over to that part of the orchard, but sadly the trees here were mired by the detritus of modern village life—ubiquitous car parts and cans of petrol perched on a fence. We returned to the original orchard and soaked in the strong green feeling of decades and centuries that had passed by. Maria brought over a book of local history. I read through some vague statements before a sentence caught my eye: "The story goes that they [the Spanish] found a huge spring and remnants of an apple orchard left by Indians centuries ago, with one tree five feet in diameter. They named the town Manzano."

Before leaving Manzano, we knocked at Maria's door and gifted her a glass jar of pecans. In New Mexico culture, old-timers delight in sharing food with neighbors and friends, and this tradition resonates with the culture I grew up in. Maria had been watching TV, and strains of a song wafted out of the open door. She smiled, pleased by the pecans—she would bake a cake for her family with them—and urged us to return later in September when the apples would be at their peak. Her grandson had told us that they grew four varieties but he didn't know the names.

"My husband and I moved in here fifty years ago," Maria said. "Before that my aunt and uncle lived here. The place has been in the family forever."

"The orchard is beautiful," I said.

"I love it," she smiled. "And my children have been happy here. Thanks be to Jesus."

As we walked away, she called after us to "come back."

We got into the car and onto the quiet road with a bagful of apples from what may be the oldest private orchard in the country. Few know of its significance, or the stories that invigorate the apples rolling about in our trunk. A landscape of bluish mountains and tawny fields flashed before us as we headed homeward to Santa Fe. Over the last centuries, the same landscape had greeted the Indigenous peoples and Spaniards countless times. They had planted apple seeds in Manzano more than two hundred years back and, the next morning, we would eat their zesty fruits for breakfast.

Chapter Twenty-one

Newtown

OCTOBER

Author and Nicholas Nehring, fruit gardener, at the South Orchard, Monticello. Photo by Roger Gettig.

By early fall, it was clear that my apple pilgrimage in Kazakhstan had not been approved—and that I may not find what I was searching for in the fragmented forests anyway. But a new opportunity came my way—to enter the seedpod that gave birth to America. I would live for a month in near solitude at Tufton Farm, near the base of Monticello (Little Mountain), and do research at the adjacent Center for Historic Plants. If I were to peer out from my cottage, I could see the structure where Thomas Jefferson's grandson once lived with his family. Jefferson may be a flawed man, but he is one of the focal points of the country's history and philosophy. He also loved apples and maintained that no apple in Paris could hold a candle to the Newtown Pippin. One October morning, I found myself acquiring a stock of Newtown Pippins at a Charlottesville farmers' market. A woman from the Carter Mountain Orchard confirmed that the pale-green apple is from the 1700s. The Newtown is one of the oldest-known cultivated apples in the country.

"We prefer to call it the Albemarle Pippin," Anna Berkes at the Jefferson Library told me. She was referring to a Virginia variant of the Newtown Pippin. In 1777, when Colonel Thomas Walker returned home to Virginia after fighting the battle of Brandywine under George Washington's command, he brought back scions from the village of Newtown on Long Island. Jefferson was already planting apples at the time. From the late 1760s to 1814, over a period of some fifty years, Jefferson and his enslaved workers planted over a thousand fruit trees in the North and South Orchards cresting Monticello. This may seem to be an extraordinary number of fruit trees, but many didn't survive and had to be replaced. Born in 1743, Jefferson grew up in a Virginia culture of the farm orchard; apple and even peach orchards were regular features that enhanced the value of ordinary farms. It was fashionable among plantation owners to cultivate a fruit garden styled along the lines of the British garden. Interestingly, Jefferson and Washington wanted to stamp a fresh American signature on their gardens and orchards while consulting books by British horticulturists such as Philip Miller, who had firmly slotted the apple into the *Malus* genus, and whose *Gardener's Dictionary* was a hit in America.

Jefferson began planning out an apple orchard even before construction was underway on his Monticello house, and he would go on to grow eighteen apple varieties in his two orchards. As I researched these old varieties, the specter of Jefferson's enslaved people also hung over me. The matter came to a head when my family visited and I took Pika to an exhibit hall to show her the original Declaration of Independence. I told her that Jefferson was the author of a key document that articulated "the colonists' right to revolution" but also signals to American citizens, including her, the nature of our own rights: "We hold these truths to be self-evident, that all men are created equal, that they are endowed by their Creator with certain unalienable rights, that among these are Life, Liberty and the Pursuit of happiness."

"People didn't always have these rights," I cautioned.

"Did Jefferson have slaves?" she asked bluntly.

"Sadly, yes. Even though he didn't believe in slavery, and even abhorred it."

"Then why didn't he release his slaves?" she asked, with wide-eyed shock.

The clarity with which nine-year-old Pika asked the question struck me. I could all but see her mind whirring: How can a man own human beings when he believes it to be wrong? Over the next few days, she and I engaged in a discussion that painted a portrait of Thomas Jefferson in his time. We began with the admission that Jefferson acted wrongly and against his conviction in the matter of slavery. He epitomizes an inherent contradiction at the heart of the country's birth: a cry for liberty from men who stood on the backs of enslaved workers. Jefferson personally was in serious debt and could not see an economic way out of slavery. Conveniently, he came to believe that slavery was a problem for generations after him to solve. The irony is that he owed $107,000 (some $3 million today) when he died, and it took his grandson, Thomas Jefferson Randolph, a lifetime and then some to pay off the debt. The moral debt is a separate issue.

History can become relevant when we drink from the past, even its bitter aspects, and consider how we might walk forward. How can we look our children in the eye and explain the cruelties that once prevailed? Jefferson is a characteristic figure: We can learn from his fervent enthusiasm for plants and trees while also growing clear-eyed about how the foundations of our country—structural, agricultural, and economic—were built by enslaved labor.

Jefferson liked to sow seeds in his vegetable gardens, especially peas, lettuce, and cabbage ("Next yr, we'll sow our cabbages together. That'll be a happy time," he wrote his daughter Martha), but he relied on enslaved people to do the heavy labor in the orchard. There was Jupiter (his last name was likely Evans), whom Jefferson all but grew up with and who was his trusted attendant when he was a young man at the College of William & Mary. In July 1774, "when illness forced Jefferson to turn back on his way to the Virginia Convention," he entrusted Jupiter to carry "copies of what became *A Summary View of the Rights of British America*. Jupiter carried to Williamsburg the words: 'The abolition of domestic slavery is the great object of desire in those colonies where it was unhappily introduced in their infant state.'" It was again Jupiter who later carried apple trees to Poplar Forest—a retreat that Jefferson's wife, Martha, had inherited from her father—to set an orchard there.

Ursula Granger, an enslaved woman originally from Martha's household (Jefferson avoided buying and selling slaves unless it was done to reunite families, but in this case, his wife asked him to make the purchase), would become the Jeffersons' cider maker. In addition to nursing the Jeffersons' daughter, also Martha, Ursula may have been a maternal figure for the child after her mother died. Jefferson held that Ursula was "the only person 'who unites trust and skill' for superintending the annual bottling of the cider."

The main apple variety that Jefferson grew for cider was the Hewes Crab, a Virginia variety that is a cross between an apple and a crabapple. The Hewes Crab matures in August, and when I walked through the North Orchard in October and wound my way through its fifty-one trees, the small crop had already been harvested to be pressed for cider. The few leaves left on the trees were crisping, and the orchard looked forlorn. The late afternoon light was radiant, however, and I could imagine how enticing the orchard might have once been.

Over a decade back, Albemarle Ciderworks pressed a small batch of Hewes Crab cider from the apples in Jefferson's North Orchard, and they were hooked. It was a small crop of apples, but the cider was exquisite. Charlotte Shelton, a co-owner of Ciderworks, situated some eight miles away from Monticello, reached out to an orchardist to ask if he could grow the apples in the quantities she needed. The orchardist pointed out that the Hewes Crab is a small apple and thus a lot of work to harvest.

"I know that," Charlotte said. "How much would you need?"

Charlotte's response to the grower's reluctance is reflective of her personality: She has a dash of old-world charm, solid knowhow, and hard pragmatism.

"Thirty-five dollars a bushel," the grower said.

"Fine," Charlotte said. She knew that growers get fifteen dollars a bushel for dessert apples, but she had a hunch that the investment would be worthwhile. The Hewes Crab has a strong amount of sorbitol, a naturally occurring sugar alcohol, which gives its cider a rounder, fuller mouthful of sweetness. Charlotte enthusiastically brought me her award-winning Hewes cider, and I sipped the drink, tasting its legendary notes of cinnamon.

The cidery now grows roughly fifty to seventy-five Hewes Crab trees, and I had hoped to walk through the orchard with Charlotte, but a cold drizzle had picked up and I was loath to ask her to get drenched for my sake. As the softly gold cider trickled down my throat, I told myself that this is what Jefferson would have been pressing from his Hewes Crabs.

"Nothing is more desired generally than fine Hughes' crab cyder," Jefferson wrote on February 2, 1802, when ordering a year's supply: ten barrels for Washington and four to be sent to Monticello. Still, his favorite cider apple was another Virginia variety called Taliaferro or Robinson's, which is now lost. In addition to the bottling that Ursula Granger did at Monticello, Jefferson ordered copious amounts of cider. His daughter, Martha, and her large troupe of children had settled in Monticello, where Martha kept house for her father, and a near constant flow of visitors dined here in accordance with Virginia hospitality. Ten years later, on April 6, 1812, Jefferson paid "Charles Massey for cyder at 10½ d pr. gallon." He would continue "to purchase his cider supply from Charles Massie (d. 1817), whose orchard at Spring

Valley on the southwestern border of Albemarle County later became renowned for its Albemarle Pippins."

During my time at Tufton Farm, I acquired a treasure-house of apple varieties from Albemarle Ciderworks and other orchards. On a sultry autumn day, the Ashmead's Kernel, a British variety from the eighteenth century, was genuinely refreshing. The skin uncannily resembles a pear, but the fruit is small and round. The Ashmead has a subtle, complex flavor with a custardy undercurrent; my mouth was left with a surprising memory of mint. The Virginia Beauty, discovered as a chance seedling in 1826, was so fragrant that I might well have been consuming fragrance, and the firm skin of the 1870 variety, Arkansas Black—deepest red, approaching burgundy—was gorgeous to behold. On another visit I picked up the White Winter Pearmain, which has slightly tough skin but a delicate, lasting flavor. It pairs beautifully with jasmine tea; both apple and tea are harmoniously green. I didn't, however, manage to taste a dry cider with a green-apple aroma called Jupiter's Legacy, which has been called the "quintessential Virginia cider" and whose name seems to honor the legacy of Jupiter Evans.

Monticello's South Orchard sprawls over roughly seven acres and once had a rich variety of fruit trees. In 1812, Jefferson was growing four hundred fruit trees, including apple, peach, and plum. Walking in the orchard, I saw traces of Bailey's Way, a path named after Robert Bailey, a Scottish gardener who worked here between 1794 and 1796. Eighty-four fruit trees grow in the South Orchard today. This summer, Nicholas Nehring, the new fruit gardener, had been shoveling grass from the base of a row of peach trees, in preparation to mulch them, when he came upon "bits and pieces," glass and ceramic, what he calls "historic debris." I fingered a substantial remnant of what might have been a wine bottle made of thick dark glass. Past archaeological digs in the orchard found remnants of a stone wall beyond the peach trees— originally built by enslaved people over a period of three years. While the process of restoring the wall was completed in the early 1980s, the restoration of the orchard is an ongoing and complicated endeavor. For the last five years, the grounds lacked a fruit gardener. The base of one Albemarle Pippin tree is covered with poison ivy, and another had been smothered by grape vines. This past July, the first week Nicholas started here, he cut back the grape vines only to watch as the apple leaves, which had grown used to shade, got fried.

Jefferson faced similar frustrations. He spent long periods away from Monticello—as a diplomat in Paris, secretary of state and vice president in Philadelphia, and a two-term president in Washington—and invariably returned to find his orchard in some form of disarray. Many of the fine Mediterranean and European fruit trees he had acquired with great enthusiasm didn't thrive, whereas some apple varieties, notably ones that had originated in America, persisted. Eventually, in what I see as the most pragmatic stage of his apple love, he would focus on the hardier varieties.

One hardy fig variety was *Ficus Carica "Marseille,"* the Marseille fig. Jefferson planted these favorites in "submural beds," as he called them, against the south-facing stone wall—so that the shrubs would be protected from colder weather. New plantings of Marseille figs have been growing here since the 1990s, and a red fox has dug a hole in the center of a huge thriving fig shrub. The first fig I tasted was a bit watery—maybe due to two weeks of violent September rains. The second was better, delighting me as the greenish-purple skin gave way to pink flesh inside—I flicked off an ant and continued to eat. Nicholas and I peered into the fox's hole and saw no fox, but on another day, I saw evidence of chewed labels. Peggy Cornett, curator of plants at Monticello, told me that the fox likes to chew wooden plant labels. "It's insult to the injury," she added. The summer has been hard, with deer consuming parts of the vegetable garden: the okra bed sports hoofprints, and the artichokes are devastated. A fawn skull was recently found in the vegetable bed, and Nicholas is not mourning the fawn.

"A coyote probably got it," he said casually.

Unlike the deer, the red fox "pays her way." She catches voles who might otherwise damage the roots of the apple trees. "I haven't seen voles here lately," Nicholas remarked. What we did see, on another visit, was fox scat, wooly gray, at the base of an apple tree. The fox especially likes peaches. Nicholas showed me a handful of peach seeds at the base of a tree; as usual, the fox had left them scrupulously clean, in a neat pile. She only eats the good peaches and spits out the rotten ones. One day, another grounds person saw the fox dancing around a groundhog until she finally managed to give it a bite and then take it down.

On sun-splashed afternoons, I initiated discussions with the grounds people about the wildlife who can be our allies in the orchard. Wooden fenceposts are gridded in the center of the orchard, where Muscadine grapes grow. One fencepost has a solid hole drilled in, and Nicholas has seen bluebirds come in and out of the hole. "I've enjoyed the bluebirds," he said.

As three or four bluebirds arced through the warm air, we spoke about how bluebirds can be beneficial, and I once again recounted Gordon Tooley's maxim: "Let the bluebird do the morning shift, and let the bats take over at night."

"And let swifts do the afternoon shift," the director of horticulture added.

What followed was an excited discussion about how to invite bats to the orchard. Our afternoons in the orchard had a community-making flavor to them.

Several buckeyes and an orange sulphur butterfly floated by, pausing on occasion over thick grass, and later a migrating monarch would waft by. Another encouraging sign was a North American wheel bug, an insect one entomologist has called "the lion or eagle of your food web." A mockingbird serenaded me most times when I was in the orchard; the bird was Jefferson's beloved pet, and the bird's descendants maintain a strong presence in the area. The cheerful song was good medicine before coming upon a struggling Esopus Spitzenburg, an eighteenth-century New York apple favored by Jefferson for dessert and still considered one of the finest varieties ever known. Jefferson planted thirty-two Esopus Spitzenburgs in the South Orchard between 1807 and 1817. The trunk of the tree before us sported a "sewing-needle eye."

"Some bad pruning decisions were made here," Nicholas said. "It has some fungal stuff blasting out of the bark. It'll come back but it's not long for this world."

The Esopus Spitzenburg variety came from north of here, and the warming climate could have made it less suited to Monticello's microclimate. The vagaries of climate have intensified the orchard's struggles.

"We're having extended dry periods most years," Peggy told me. A storm that came at the heels of Hurricane Debby took down a peach tree this summer. The summer's drought was followed by two weeks of violent rains in September, when Hurricane Helene slammed neighboring North Carolina.

Before weather collection became institutionalized, weather data was collected primarily by citizen scientists. Researchers at Princeton University, where Jefferson's weather and climate records are archived, note that between July 1776 and June 1826, Jefferson made "19,000 temperature readings and observations of sky conditions, periodic runs of pressure, humidity, wind direction and force, precipitation measurements, and phenological observations, or notes on the seasonal changes seen in wildlife and plant growth. . . ." His notes and measurements are believed to form a "valuable resource for reconstructing the climate of the early American Republic." Jefferson had a scientific mind, and were he alive today, he would have almost certainly paid attention to the existential challenges climate change is thrusting our way.

In addition to the weather, Jefferson kept an eye on the dimensions of his garden. On March 31, 1774, he wrote in his *Garden Book*: "laid off ground to be levelled for a future garden. the upper side is 44. f. below the upper edge of the Round-about and parallel thereto. it is 669. feet long, 80 f. wide, and at each end forms a triangle, rectangular & isosceles, of which the legs are 80. f. & the hypothenuse 113. feet." In the same entry, he noted that he had planted twenty-four apple trees and nineteen cherry trees, both from the Mountain Plains, "the plantation of Michael Woods on Mechum's River and Lickinghole" in Albemarle County.

Note that he pens down only how many apple or cherry trees he planted and doesn't identify them any further. Four years later, in March of 1778, when Jefferson grafted and planted new apple trees in the orchard, he took care to note the varieties: Newtown Pippins, Medlar Russetins (*Mespilus germanica* or medlar is "not included in the apple group today"), Golden Wildings (a medium yellow apple from North Carolina), Robinson (also known as Taliaferro), Codlin ("a favorite apple in England for pies and stewing"), White (an eastern

Virginia apple, which Downing characterized as having tender white flesh, on the verge of melting). At the same time, Jefferson planted apple seeds in his nursery. His apple love was deepening.

In the years that followed, when Jefferson's travels took him from Paris to Philadelphia to Washington, there is a refrain in his correspondence of yearning to be back at Monticello to attend to his fruit trees and the task of growing them. The apple was not his only love; during his Paris years, he was smitten with the olive tree. After retirement from presidency, he was obsessed with the peach. The contours of Jefferson's arboreal obsession are complex, but what is clear from bursts of correspondence, boxes of plants sent and received, apples shipped and cider ordered is that this love animated his friendships with a host of relations, colleagues, and fellow fruit growers.

On October 26, 1805, Jefferson was in Washington and wrote to Madame de Tessé in France, describing the contents of a seed box, "4 feet long, and 1 foot wide and deep," which he was sending her despite the blockade of Havre and "the irregularities committed on the ocean by the armed vessels of all the belligerent powers that nothing is safe committed to that element." In a footnote, he added: "since writing the above I have been able to get some of the Pyrus coronaria, or malus sylvestris virginiana floribus odoratis of Clayton, both the blossom and apple are of the finest perfume, and the apple is the best of all possible burnishers for brass and steel furniture which has contracted rust." Characteristically, in one snippet of a letter, written in haste, Jefferson dispensed opinions on politics, botany, and even housekeeping. He was, of course, enthusiastic about all of these subjects, but through his relentless horticultural experimentation, it was as though he were trying to grow something new—a uniquely American way of being.

After his retirement from public life, Jefferson hoped to devote himself to his vegetable garden and orchards. While he planted many new trees during this time, the penultimate stage of Jefferson's career as an orchardist seems to have been characterized by detachment. In correspondence, he began to complain about how his orchard was decaying. He had gone through several ups and downs with more than a thousand fruit trees over four decades and became almost philosophical about the vagaries and disappointments that came his way. Which may sound familiar to many orchardists.

I was not surprised to discover that Jefferson was a walker, and he recommended this exercise to family and friends ("of all the exercises walking is best. . . . Not less than two hours a day should be devoted to exercise, and the weather should be little regarded.") A walker can closely observe the natural world, its trees and plants, and how the weather changes from one day to the next. For most of the month when I lived at Tufton Farm, his grandson's estate where the Center for Historic Plants is now based, I didn't rent a car and instead walked extensively in the area. But only in natural areas, for cars whizzed by so fast on the Jefferson Parkway that more than one person advised me not to walk along its almost nonexistent shoulder. I walked across vast grassy fields, many of which are now leased to cattle ranchers, and over hilly meadows rimmed with towering trees. Once, I came upon two ranchers who asked me to watch where I stepped; good advice, since cow dung was embedded in the grass, and on dewy mornings it was enough that my shoes and socks were soaked through, I didn't also want to step into cow patties. Eventually, the fields spilled me out onto a path crunchy with fallen maple leaves. On this forested path, crickets ruled and, no matter the time of day, light seemed to be slipping. While I had felt relatively safe in the fields, anxious only about lingering ticks, I couldn't shake off a spooky feeling on the forested path. The bear sometimes seen around Monticello probably also drifts through here. Steadied by the gentle tapping of a woodpecker, I made my way to a creek of the Rivanna River. Along the bank, I was delighted to see ferns spreadeagled over the ground. I startled a deer and gasped when he broke into a run. The creek's flow was feeble, and power lines marred the view. I soon turned around.

    One afternoon, Peggy suggested that we hike all the way to the river. It was already late October, but it was sunny and hot and so dry that Peggy found it worrisome—the ground was usually much wetter in the fall. Once we reached the creek, I improvised a walking stick to help us get across. Then we scrambled up a slope, weaving our way through wild rose bushes, getting snagged on thorns, and all but crawling out of a fence, which at least wasn't barbed. We made our way to an old overgrown path that eventually led us to the Rivanna River. The river was vast, and farther along its banks were lively rapids. We stood on a stone slab, peering at the gray water. Jefferson likely fished here with his boyhood friend—and later, personal attendant—Jupiter, the lines between personal desires and societal norms blurring, as they would do for the rest of his life. This afternoon, a blue heron was fishing. From a different vantage point, we saw the heron fly away, his great bluish wings contrasting with the gray of the capacious river. Peggy pointed out a bald eagle nest on a platform, which a utility company eventually fitted into a transmission tower—to accommodate a pair of eagles who has been returning to the same spot to nest for the last twenty-two years. When I walked through fields where there were no paths, I hewed to the remnants of tire tracks in order to avoid taller grasses that might harbor ticks. I walked past the stone ruins of two historic structures. Most days, I stopped at the garden or library of the Center for Historic Plants and researched the history and living specimens of some of the oldest plants introduced to the country, especially old apple varieties: Father Abraham apple, "first mentioned in 1790 and brought to Virginia by early German settlers"; Pomme Gris, a "much admired" apple from Cincinnati—John Armstrong sent cuttings to Jefferson in 1804.

While most of my walks were solitary, I wasn't alone. I was accompanied by the cries of the blue jay, a cardinal's sudden dive, flickers undulating from one elm to another, meadowlarks singing on fenceposts before dropping to the grass. A herd of white-tailed deer watched me closely, often racing ahead. I avoided the wary looks the cattle gave me but made eye contact with two racoons thrusting pointed noses in the grass. The pair tolerated my presence for a surprisingly long time. At last, they considered me tentatively before darting into their hole and inadvertently revealing the location of their home. During the walks, I gazed at the rolling mountains, which were almost always before me, and the mature hackberry trees shimmering in magical yellows and ochre. The drought has been so persistent here that the changing of the colors, a ritual that humans have experienced since the beginning of our time on earth, was short-circuited this year. The tulip poplar leaves got fried on the trees, crisping prematurely and dropping to the ground, and many trees such as the hackberry turned color all at once, briefly, when in years past the colors would change in graceful succession. "I've never seen anything like it," Peggy said. That's something, coming from a plant curator who has worked at Monticello for over forty years.

I woke up once to a flock of wild turkeys. Another evening, while a colleague was driving me back from the orchard, we came upon a tufted titmouse hovering near a red-bellied woodpecker eating his fill of ants, smack in the middle of a path. The moments spent with wildlife filled me up, as did walking through fields, limbs aching. One evening, I sensed viscerally that I was seeing the same river and mountains as people who walked through here as far back as three hundred years ago, and more. This area is the ancestral homeland of the Monacan people, who continued to honor their burial mounds here as late as the mid-1700s.

One morning in mid-October, Peggy and I walked to an ancient feral apple tree, which she'd last seen growing in the area some ten to fifteen years back. As we searched the edge of a field for the apple tree, my shoes and socks got wet as rags, and my pants were rimmed with red clay. But the tree was gone. It was disappointing to think that it was maybe cut down or had fallen in a storm. After Jefferson's grandson, the land was sold to a family who grew an orchard here, and the old apple tree was likely a relic from that orchard. As I kept on walking, I thought about how some of the mature oaks, elms, and maples that were browning were descendants of trees from the time of our country's birth. That time was rife with contradictions as, in different ways, is our own. Befriending these multitudes of trees, I experienced rootedness to the land. Light fell across the landscape in ways that enveloped the trees and grass and touched my face like a benediction.

We find ourselves at an unprecedented crossroads: We must commit to healing nature if we mean to live in a world where nature herself will not become an overwhelmingly destructive force. Embraced by light and a feeling of oneness, I kept on walking, ever deeper on green and gold paths, until one day when a chill permeated the evening air. The glorious store of apples in my kitchen was dwindling, and I gifted the last two bags to friends. It was time to return home, even as I experienced a heightened awareness that apple culture has expanded my sense of home. I will not walk such verdant paths in Santa Fe nor look upon such giant trees, but I will soon inhale fragrant smoke from piñon fires and in the coming months walk in snowy forests. Sometime during the winter, I will dream of Virginia apples and pray that their scent might one day call me back.

Epilogue

Author's father plants a tree outside their home in the foothills of the Himalayas.

The packet of letters had all but fallen on Michael as he was looking for my passport. When a trip to the national park in Kazakhstan was still a possibility, we wanted to be ready in case my extended trip was approved. A family member must have brought the letters along with some other things and handed them to me wordlessly. I had put the tawny eight-by-eleven-inch envelope in a safe place, up in a closet, meaning to take a closer look later. But extended family was visiting at the time, and I somehow forgot about the envelope.

I pulled out the cache of letters. My eye fell on one, then another. How familiar those blue sheets looked! As I glanced at the third letter in the stack, I could scarcely believe what I was reading. An entire letter devoted to apples! So, I hadn't been imagining the visit to the orchard? Such tears blurred my vision that it made it difficult to read. The letter was from the orchardist. He was writing to my father, but first he thanked the children (me and my older brother) for our letters, in which we'd inquired about the apples. I looked at the date. I was four years old at the time, and I was already corresponding with an orchardist? A glimmer of a smile, a gasp. Blotting my tears, I read a little more. Sadly, the apples are not yet ready, he wrote. That is why I didn't write to you earlier. Then he addressed my father, discussing with him the challenging peculiarities of that apple season. I was wracked with quiet sobs for some reason and gently put away the fragile letter. That weekend, I purchased an archival box into which I emptied the contents of the weathered envelope. There are stories in there for

The author as a young child.

another day. It feels enough to know that when I was four, I experienced the rhythms of nature, the rasa of apples, and it all made a deep enough impression that I wrote to an orchardist asking when I might come back.

Today, when I see Mia climbing apple trees without a blink of hesitation, I wonder at the mysterious ways in which traditions get passed on. At the moment, she sports chocolate-colored scrapes on her legs from when she fell out of a tree a few days back. She had climbed an old cottonwood with a deeply slanted trunk that hews the Santa Fe River, and there was no good way to climb down. The textbook cat dilemma.

Mia came down with the intuitive precision of a veteran but lost her footing toward the end, letting out a yelp as she descended in free fall. She picked herself up and walked away with nasty bleeding scrapes all over her legs (yes, she was in shorts—and flip-flops).

The ways in which my girls embrace trees reminds me to go on embracing nature. Alas, we don't live nestled against wild areas, but we follow orchard trails when we can, with the trails sometimes leading us into a micro-wilderness. The riches of these in-between spaces are surprising. It is an act of love to stretch and pick the fruits you can reach. Any scrapes or even ancient wounds can heal when the rasa of wild fruit percolates the body and soul.

Wandering among apple trees, I occasionally crack the mundane surface of life and slip through this crack to live more fully. To travel still deeper, I slip away into the forest, among a grove of white pines or mature aspens. In the digital age, I will take what cracks I find to discover greener worlds, to become a bear roaming the wild. Trees hold and support our lives. Loving magical trees in the orchard and the forest has the power to regenerate us and the planet. So, we have admirable reasons to cultivate tree love. Together, you and I can move the needle on biodiversity and, as fall mellows, we can at last unwind and savor the acidic-and-honeyed fruits of our labors.

Michael and Mia watering. Photo by Priyanka Kumar.

Acknowledgments

It is fitting that I'm writing these notes on Valentine's Day since this book was fueled by love for apples and the wild (and coffee). My deepest thanks to my superb editor Rebecca Bright for her sustained enthusiasm and for driving from DC to Monticello to meet me atop the mountain.

I felt buoyed by biologist David George Haskell's excitement about the ideas I explore in the book. Sy Montgomery recognized the book's essence—thank you! Jane Smiley appreciated and supported the work at the right moment, as did Andrea Barrett. My agent Leslie Meredith is incredibly kind and intelligent and a font of wise counsel. Thanks to Lindsay Botts for discovering my work and bringing it to *Sierra Magazine*. The photographer Don Usner generously made my photographs shine before the book went into production.

Shimmering gratitude to all apple historians, scientists, and orchardists who are guardians of the land. I am fortunate to know the incomparable Gordon Tooley and his wife Margaret Yancey (whose breads are to die for). David Kenneke and master gardener Thelma Coker welcomed me to Cimarron with exquisite warmth. John Trujillo is an inspiration. Gail Haggard is a rock star. Marcy is sublime, and I am very grateful to Steve Vrooman for taking me to meet him. The lovely Carmella Padilla introduced me to Sean Paloheimo at Las Golondrinas. Elizabeth Brown took me to the Tumacácori National Historical Park and the Mission Garden where I met the dedicated orchardist Jésus García.

Stephanie Brock and her cousin Les Swindle graciously hosted me and my family at DeHaven Ranch, and botanist Bob Sivinski introduced me to Stephanie and later read the manuscript with enthusiasm. Botanist Tom Antonio doggedly tracked down the cacao story. Archaeobotanist Robert Spengler and biologist Amandine Cornille generously answered my inquiries. The historians Curt Meine and Dan Bussey graciously answered a query about a Wisconsin apple variety.

An ICJS Fellowship allowed me to do research at the Center for Historic Plants in Monticello. I am grateful to librarian Anna Berkes for suggesting apt research material and taking me on a visit to Albemarle Ciderworks, where I eventually met Charlotte Shelton. Librarian Endrina Tay pointed me to Jefferson's climate notes. Andrew Davenport and Caitlin Lawrence made all the wheels turn. Philip Herrington was the most courteous of colleagues. Hikes and meals with Peggy Cornett, curator of plants, were a highlight. Director of horticulture Roger Gettig and head gardener Nicholas Nehring were always there to share knowledge and a laugh.

At the Kachemak Bay Writers' Conference in Alaska, my fellow faculty did their best to ply me with cider oddities. It was a treat to meet Nancy Lord.

A residency at PLAYA in Oregon allowed me to research orchards in the area. Thanks to Carrie Hardison for facilitating my stay and to Kris Norris for connecting me to local orchards and photocopying valuable research material. The team at Island Press, including Sharis Simonian, Julie Marshall, Jaime Jennings, and Julie Greene, has been incredibly supportive.

Thanks to Jeff Orlowski-Yang for the way your eyes lit up when I spoke about the "micro-wild."

Deb Winslow is a lovely friend, and gave the manuscript an intelligent read when I needed it. Richard Ian Greene reliably made me laugh in the middle of a sentence. I am grateful for Peter Rainer's sparkling wit and his support. James Ragan unstintingly offered appreciation and support. Thanks to Jack and Juanita Ortega for sharing apricots and jam. It was delightful to do "apple exchanges" with Ms. Cherry, who also gave us a tour of her orchard. Peter is an engaged grandfather to the kids and often asked about the book's progress.

Radiant thanks to Mia and Pika (nicknames) for bearing with many wild excursions to areas with "no buildings" in sight. Mia began to climb apple trees at the age of four and taught the rest of us how to snag apples from trees and eat around the worms.

My loving gratitude to Michael, the best person I know, for reading drafts of the manuscript, tracking down science papers, engaging in discussions on apple science—and especially for becoming a fellow apple enthusiast.

Apple cores. Photo by Priyanka Kumar.

Select Sources

1. An Apple for Ursula

Brown, Marley. 2020. "On the Origin of Apples." *Archeology*, January/February.

Cornille, Amandine, Pierre Gladieux, Marinus J. M. Smulders, et al. 2012. "New Insight into the History of Domesticated Apple: Secondary Contribution of the European Wild Apple to the Genome of Cultivated Varieties." *PLOS Genetics* 8 (5). https://doi.org/10.1371/journal.pgen.1002703.

Coverstone, Nancy et al., ed. 2000. "Wild Apple Trees for Wildlife." In *Habitats: A Fact Sheet Series on Managing Lands for Wildlife*. University of Maine Cooperative Extension Bulletin # 7126.

Dzhangeliev, A. D. 2002. "The Wild Apple Tree of Kazakhstan." In *Horticultural Reviews: Wild Apple and Fruit Trees of Central Asia* 29. Edited by Jules Janick. Wiley.

Hinton, David, trans. 2014. *Chuang Tzu: The Inner Chapters*. Counterpoint.

Juniper, Barrie E., and David J. Mabberley. 2006. *The Story of the Apple*. Timber Press.

Nabhan, Gary Paul. 2009. *Where Our Food Comes From: Retracing Nikolay Vavilov's Quest to End Famine*. Island Press.

Rilke, Rainer Maria. 1996. *Rilke's Book of Hours: Love Poems to God*. Translated by Anita Barrows and Joanna Macy. Riverhead Books.

Spengler, Robert Nicholas. 2019. "Origins of the Apple: The Role of Megafaunal Mutualism in the Domestication of *Malus* and Rosaceous Trees." *Frontiers in Plant Science* 10. https://doi.org/10.3389/fpls.2019.00617.

Witherell, Elizabeth Hall, ed. 2001. *Henry David Thoreau: Collected Essays and Poems*. Library of America.

Wu, Rui, Fasheng Lou, Juan Yu, et al. 2024. "The Smallest Known Complete Dinosaur Fossil Eggs from the Upper Cretaceous of South China." *Historical Biology*, 1–10. https://doi.org/10.1080/08912963.2024.2409873.

2. In Search of Feral Apples

New Mexico Department of Game and Fish. 2022. "Keeping Bears Alive and You Safe." https://wildlife.dgf.nm.gov/download/keeping-bears-alive-and-you-safe/.

Williams, Gary S. 1989. *Johnny Appleseed in the Duck Creek Valley*. Johnny Appleseed Center for Creative Learning.

Witherell, Elizabeth Hall, ed. 2001. *Henry David Thoreau: Collected Essays and Poems*. Library of America.

3. The Venerable Crab

Besnard, Guillame, Jean-Frédéric Terral, and Amandine Cornille. 2018. "On the Origins and Domestication of the Olive: A Review and Perspectives." *Annals of Botany* 121 (3): 385–403. https://doi.org/10.1093/aob/mcx145.

Cornille, Amandine, Pierre Gladieux, Marinus J. M. Smulders, et al. 2012. "New Insight into the History of Domesticated Apple: Secondary Contribution of the European Wild Apple to the Genome of Cultivated Varieties." *PLOS Genetics* 8 (5). https://doi.org/10.1371/journal.pgen.1002703.

Cornille, Amandine, Ferran Antolín, Elena Garcia, et al. 2019. "A Multifaceted Overview of Apple Tree Domestication." *Trends in Plant Science* 24 (8): 770–782. https://doi.org/10.1016/j.tplants.2019.05.007.

Duan, Naibin, Yang Bai, Honghe Sun, et al. 2017. "Genome Resequencing Reveals the History of Apple and Supports a Two-Stage Model for Fruit Enlargement." *Nature Communications* 8:249. https://doi.org/10.1038/s41467-017-00336-7.

Dunbar-Wallis, Amy, Gayle M. Volk, Alexandra M. Johnson, et al. 2022. "What's in a Name? The Importance of Identity in Heirloom Apple Tree Preservation." *Plants, People, Planet* 5 (1): 39–46.

Dzhangeliev, A. D. 2002. "The Wild Apple Tree of Kazakhstan." In *Horticultural Reviews: Wild Apple and Fruit Trees of Central Asia* 29. Edited by Jules Janick. Wiley.

Ellisona, Rosemary, Jane Renfrew, Don Brothwell, and Nigel Seeley. 1978. "Some Food Offerings from Ur, Excavated by Sir Leonard Woolley, and Previously Unpublished." *Journal of Archaeological Science* 5 (2): 167–177.

Gilbert, W. M., and Ammon. 2020. "Wilson's Warbler." *Birds of the World*. ed. P. G. Rodewald. Cornell Lab of Ornithology.

Juniper, Barrie E., and David J. Mabberley. 2006. *The Story of the Apple*. Timber Press.

Korban, Schuylan S., and R. M. Skirvin. 1984. "Nomenclature of the Cultivated Apple." *HortScience* 19 (2): 177–180.

Mudge, Ken, Jules Janick, Steven Scofield, and Eliezer E. Goldschmidt. 2009. "A History of Grafting." In *Horticultural Reviews* 35, edited by Jules Janick. Wiley.

Spengler, Robert N. 2020. *Fruit from the Sands: The Silk Road Origins of the Foods We Eat*. University of California Press.

Torgrimson, John. 2009. *Fruit, Berry, and Nut Inventory*. 4th ed. Seed Savers Exchange.

Zohary, Daniel, Maria Hopf, and Ehud Weiss. 2012. *Domestication of Plants in the Old World: The Origin and Spread of Domesticated Plants in Southwest Asia, Europe, and the Mediterranean Basin*. 4th ed. Oxford University Press.

4. Seeking Celestial Apples

Couillard-Després, Azarie. 1913. "Louis Hébert Premier Colon Canadien Et Sa Famille." Paris: Desclée, De Brouwer & Co.

de Champlain, Samuel. 1859. *Narrative of a Voyage to the West Indies and Mexico in the Years 1599–1602*. Translated by Alice Wilmere. London: Hakluyt Society.

DeGraaf, Richard M., and Gretchin M. Witman. 1979. *Trees, Shrubs, and Vines for Attracting Birds*. University of Massachusetts Press.

Desloges, Yves. n.d. "Daily Life: Foodways." Canadian Museum of History. https://www.historymuseum.ca/virtual-museum-of-new-france/daily-life/foodways/.

Dunmire, William W. 2004. *Gardens of New Spain: How Mediterranean Plants and Foods Changed America*. University of Texas Press.

Ivey, James E. 1988. *In the Midst of a Loneliness: The Architectural History of the Salinas Missions*. Southwest Cultural Resources Center.

MacDowell, Laurel Sefton. 2012. *An Environmental History of Canada*. University of British Columbia Press.

Martin, Carol. 2000. *A History of Canadian Gardening*. McArthur and Company.

Watson, Burton, trans. 2000. *Po Chü-I, Selected Poems*. Columbia University Press.

5. The Fruitery

Statista. 2023. "Per Capita Consumption of Fresh Apples in the United States from 2000/01 to 2022/23." https://www.statista.com/statistics/257167/per-capita-consumption-of-fresh-apples-in-the-us/.

Bynum, Helen, and William Bynum. 2014. *Remarkable Plants That Shape Our World*. University of Chicago Press.

Dolan, Susan. 2009. *Fruitful Legacy: A Historic Context of Orchards in the United States*. National Park Service.

Martin, Peter. 1991. *The Pleasure Gardens of Virginia: From Jamestown to Jefferson*. Princeton University Press.

George Washington's Mount Vernon. n.d. "George." https://www.mountvernon.org/library/digitalhistory/digital-encyclopedia/article/george#1.

McCullough, David. 2001. *John Adams*. Simon and Schuster.

Monticello. n.d. "Fruit Gardens." https://www.monticello.org/house-gardens/farms-gardens/fruit-gardens/.

National Archives. n.d. "About the Papers of George Washington." Founders Online. https://founders.archives.gov/about/Washington.

Rose, Mary. 2016. "How Native Farmers Shaped the Northwest Apple Industry." Confluence Library. https://www.confluenceproject.org/library-post/how-native-farmers-shaped-the-northwest-apple-industry-part-1-origins/.

Torgrimson, John. 2009. *Fruit, Berry, and Nut Inventory*. 4th ed. Seed Savers Exchange.

Washington, George. 1793. Letter to Anthony Whitting. May 12. Founders Online. National Archives. https://founders.archives.gov/documents/Washington/05-12-02-0451.

Wikipedia. n.d. "Thirteen Colonies." https://en.wikipedia.org/wiki/Thirteen_Colonies.

Wulf, Andrea. 2012. *Founding Gardeners: The Revolutionary Generation, Nature, and the Shaping of the American Nation*. Vintage Books.

Young, Allen M. 2007. *The Chocolate Tree: A Natural History of Cacao*. University Press of Florida.

6. Ten Thousand Varieties

Antonelli, Alexandre. 2022. *The Hidden Universe: Adventures in Biodiversity*. University of Chicago Press.

Cornille, Amandine, Tatiana Giraud, Marinus J. M. Smulders, Isabel Roldán-Ruiz, and Pierre Gladieux. 2014. "The Domestication and Evolutionary Ecology of Apples." *Trends in Genetics* 30 (2): 57–65.

Gross, Briana L., A.D. Hank, C.M. Richards, G. Fazio, and G. Volk. 2014. "Genetic Diversity in *Malus X domestica* (Rosaceae) through Time in Response to Domestication." *American Journal of Botany* 101 (10): 1770–1779.

Jacobsen, Rowan. 2014. *Apples of Uncommon Character: 123 Heirlooms, Modern Classics, and Little-Known Wonders*. Bloomsbury.

Juniper, Barrie E., and David J. Mabberley. 2006. *The Story of the Apple*. Timber Press.

Means, Howard. 2011. *Johnny Appleseed: The Man, the Myth, the American Story*. Simon and Schuster.

Nabhan, Gary Paul, ed. 2010. *Forgotten Fruits Manual and Manifesto: Apples*. Renewing America's Food Traditions (RAFT) Alliance.

Torgrimson, John. 2009. *Fruit, Berry, and Nut Inventory*. 4th ed. Seed Savers Exchange.

7. The Flow of Energy

Allen, Elizabeth Hightower, ed. 2022. *First and Wildest: The Gila at 100*. Torrey House Press.

Cornille, Amandine, Tatiana Giraud, Marinus J. M. Smulders, Isabel Roldán-Ruiz, and Pierre Gladieux. 2014. "The Domestication and Evolutionary Ecology of Apples." *Trends in Genetics* 30 (2): 57–65.

Cornille, Amandine, Pierre Gladieux, Marinus J. M. Smulders, et al. 2012. "New Insight into the History of Domesticated Apple: Secondary Contribution of the European Wild Apple to the Genome of Cultivated Varieties." *PLOS Genetics* 8 (5). https://doi.org/10.1371/journal.pgen.1002703.

Cornille, Amandine, Ferran Antolín, Elena Garcia, et al. 2019. "A Multifaceted Overview of Apple Tree Domestication." *Trends in Plant Science* 24 (8): 770–782. https://doi.org/10.1016/j.tplants.2019.05.007.

Duan, Naibin, Yang Bai, Honghe Sun, et al. 2017. "Genome Resequencing Reveals the History of Apple and Supports a Two-Stage Model for Fruit Enlargement." *Nature Communications* 8: 249. https://doi.org/10.1038/s41467-017-00336-7.

Idowu, John. 2022. "Soil Structure." Lecture. New Mexico State University.

Leopold, Aldo. 2013. *Aldo Leopold: A Sand County Almanac & Other Writings on Ecology and Conservation*. Edited by Curt Meine. Library of America.

Nabhan, Gary Paul, ed. 2010. *Forgotten Fruits Manual and Manifesto: Apples*. Renewing America's Food Traditions (RAFT) Alliance.

Preston, Shawn, Kirti Rajagopalan, Matthew Yourek, Lee Kalcsits, and Deepti Singh. 2024. "Changing Climate Risks for High-Value Tree Fruit Production Across the United States." *Environmental Research Letters* 19:124092. https://doi.org/10.1088/1748-9326/ad90f4.

Rochette, P., G. Bélanger, Y. Castonguay, A. Bootsma, and D. Mongrain. 2004. "Climate Change and Winter Damage to Fruit Trees in Eastern Canada." *Canadian Journal of Plant Science* 84 (4): 1113–1125.

Sun, Xuepeng, Chen Jiao, Heidi Schwaninger, et al. 2020. "Phased Diploid Genome Assemblies and Pan-Genomes Provide Insights into the Genetic History of Apple Domestication." *Nature Genetics* 52:1423–1432.

8. Pulling the Crank

Hesse, Hermann. 1972. *Wandering*. Noonday Press.

Lopez, Barry, and Debra Gwartney, eds. 2013. *Home Ground: A Guide to the American Landscape*. Trinity University Press.

Ripple, William J., Christopher Wolf, Jillian W. Gregg, et al. 2024. "The 2024 State of the Climate Report: Perilous Times on Planet Earth." *BioScience* 74 (12): 812–824.

9. Wild Horses

Armstrong, Ruth W. 1981. *The Chases of Cimarron*. New Mexico Stockman.

Cunningham, Matt. 2023. "The Sheep's Nose Apple." Minnetonka Orchards. April 19. https://minnetonkaorchards.com/sheeps-nose-apple/.

Dolan, Susan. 2009. *Fruitful Legacy: A Historic Context of Orchards in the United States*. National Park Service.

Dunbar-Wallis, Amy, Gayle M. Volk, Alexandra M. Johnson, et al. 2022. "What's in a Name? The Importance of Identity in Heirloom Apple Tree Preservation." *Plants, People, Planet* 5 (1): 39–46.

Jacobsen, Rowan. 2014. *Apples of Uncommon Character: 123 Heirlooms, Modern Classics, and Little-Known Wonders*. Bloomsbury.

Lifson, Amy. 2009. "Ben-Hur: The Book That Shook the World." *Humanities* 30, no. 6 (November/December). https://www.neh.gov/humanities/2009/novemberdecember/feature/ben-hur-the-book-shook-the-world.

Mitchell, Stephen, ed. and trans. 1989. *The Selected Poetry of Rainer Maria Rilke*. Vintage International.

Nabhan, Gary Paul. 2009. *Where Our Food Comes From: Retracing Nikolay Vavilov's Quest to End Famine*. Island Press.

Witherell, Elizabeth Hall, ed. 2001. *Henry David Thoreau: Collected Essays and Poems*. Library of America.

10. Industrial Fruit

Arellano, Juan Estevan. 2014. *Enduring Acequias: Wisdom of the Land, Knowledge of the Water*. University of New Mexico Press.

Environmental Working Group. 2019. "Apples Doused with Chemical After Harvest." March 20. https://www.ewg.org/foodnews/apples.php.

Espino-Díaz, Miguel, David Roberto Sepúlveda, Gustavo González-Aguilar, and Guadalupe I. Olivas. 2016. "Biochemistry of Apple Aroma: A Review." *Food Technology and Biotechnology* 54 (4): 375–397.

Gelles, David, and Vivian Nereim. 2023. "For Brazil, 'the Climate Emergency Is Already a Reality.'" *The New York Times*, December 1.

IPCC Sixth Assessment Report. 2022. *Climate Change 2022: Impacts, Adaptations, and Vulnerabilities*. https://www.ipcc.ch/report/ar6/wg2/.

National Toxicology Program. 2016. *Report on Carcinogens: N-nitrosamines*. 15th ed. Department of Health and Human Services. ntp.niehs.nih.gov/ntp/roc/content/profiles/nitrosamines.pdf.

Parker, Najja. 2019. "This Is the Healthiest Part of the Apple, According to Study." August 5. https://phys.org/news/2019-08-healthiest-apple.html.

Plumer, Brad, Raymond Zhong, and Lisa Friedman. 2022. "Time Is Running Out to Avert a Harrowing Future, Climate Panel Warns." *The New York Times*, February 28.

Sengupta, Somini. 2024. "Drought Touches a Quarter of Humanity, U.N. Says, Disrupting Lives Globally." *The New York Times*, January 11.

US Department of Agriculture Natural Resources Conservation Service. n.d. "Soil Health." https://www.nrcs.usda.gov/conservation-basics/natural-resource-concerns/soils/soil-health.

Wilson, E. O. 2016. *Half-Earth: Our Planet's Fight for Life*. W. W. Norton.

Yepsen, Roger. 1994. *Apples*. W. W. Norton.

11. The Archbishop's Garden

Hesse, Hermann. 1972. *Wandering*. Noonday Press.

Horgan, Paul. 1975. *Lamy of Santa Fe: His Life and Times*. Farrar, Straus and Giroux.

Johnson, George. 2006. "Archbishop Lamy's Parking Lot." *Santa Fe Review*, July 9.

Santa Fe New Mexican. 2021. "A Tree Fell in Santa Fe, and a City Says Goodbye." Our View editorial. December 16.

Rilke, Rainer Maria. 1985. *Letters on Cézanne*. Translated by Joel Agee. Fromm International Publishing Corporation.

Stinchcombe-Gillies, Kate. n.d. "Along the Waterways: 8 Utterly Delicious Fruits from Southwest France." French Waterways. https://www.french-waterways.com/fruits-southwest-france/.

Torgrimson, John. 2009. *Fruit, Berry, and Nut Inventory*. 4th ed. Seed Savers Exchange.

12. Community Making

Arellano, Juan Estevan. 2014. *Enduring Acequias: Wisdom of the Land, Knowledge of the Water*. University of New Mexico Press.

Mission Garden. https://www.missiongarden.org/.

Nabhan, Gary Paul, ed. 2010. *Forgotten Fruits Manual and Manifesto: Apples*. Renewing America's Food Traditions (RAFT) Alliance.

National Park Service. "Kino Heritage Fruit Trees." https://www.nps.gov/articles/000/kino-heritage-fruit-trees.htm.

13. Mapping the Orchard

Nabhan, Gary Paul, ed. 2010. *Forgotten Fruits Manual and Manifesto: Apples*. Renewing America's Food Traditions (RAFT) Alliance.

14. The Generosity of Blossoms

Browning, Frank. 1998. *Apples: The Story of the Fruit of Temptation*. North Point Press.

Cornille, Amandine, Pierre Gladieux, and Tatiana Giraud. 2013. "Crop-to-Wild Gene Flow and Spatial Genetic Structure in the Closest Wild Relatives of the Cultivated Apple." *Evolutionary Application* 6 (5): 737–748. https://doi.org/10.1111/eva.12059.

Danforth, Bryan N., Robert L. Minckley, and John L. Neff. 2019. *The Solitary Bees: Biology, Evolution, Conservation*. Princeton University Press.

Dzhangeliev, A. D. 2002. "The Wild Apple Tree of Kazakhstan." In *Horticultural Reviews: Wild Apple and Fruit Trees of Central Asia* 29. Edited by Jules Janick. Wiley.

Epictetus. 1865. *The Works of Epictetus: Consisting of His Discourses, in Four Books, the Enchiridion, and Fragments*. Little, Brown, and Company.

Evans, Howard Ensign, and Mary Jane West-Eberhard. 1973. *The Wasps*. David and Charles.

Flynt, Diane. 2023. *Wild, Tamed, Lost, Revived: The Surprising Story of Apples in the South*. University of North Carolina Press.

Gladieux, Pierre, Xiu-Guo Zhang, Isabel Róldan-Ruiz, et al. 2010. "Evolution of the Population Structures of *Venturia inaequalis*, the Apple Scab Fungus, Associated with the Domestication of Its Host." *Molecular Ecology* 19(4): 658–674.

Haskell, David George. 2012. *The Forest Unseen: A Year's Watch in Nature*. Penguin.

Jacobsen, Rowan. 2014. *Apples of Uncommon Character: 123 Heirlooms, Modern Classics, and Little-Known Wonders*. Bloomsbury.

Jahed, Khalil R., and Peter M. Hirst. 2017. "Pollen Tube Growth and Fruit Set in Apple." *HortScience* 52 (8): 1054–1059. https://doi.org/10.21273/HORTSCI11511-16.

McGrew, Roderick E. 1985. *Encyclopedia of Medical History*. McGraw-Hill.

Miller, Cristanne, and Domhnall Mitchell, ed. 2024. *The Letters of Emily Dickinson*. Belknap Press.

Minelli, Alessandro. 2018. *Plant Evolutionary Developmental Biology*. Cambridge University Press.

Nabhan, Gary Paul. 2009. *Where Our Food Comes From: Retracing Nikolay Vavilov's Quest to End Famine*. Island Press.

Rondeau, Sabrina. 2024. "Digging Below the Surface: Hidden Risks for Ground-Nesting Bees." *Science* 386 (6723): 739-A–739-B.

15. Cow Creek Runs Through Paradise

Jacobsen, Rowan. 2014. *Apples of Uncommon Character: 123 Heirlooms, Modern Classics, and Little-Known Wonders*. Bloomsbury.

Luby, J., P. Forsline, H. Aldwinckle, V. Bus, and M. Geibel. 2001. "Silk Road Apples—Collection, Evaluation, and Utilization of *Malus sieversii* from Central Asia." *HortScience* 36(2): 225–231.

Nikiforova, Svetlana V., Duccio Cavalieri, Ricardo Velasco, and Vadim Goremykin. 2013. "Phylogenetic Analysis of 47 Chloroplast Genomes Clarifies the Contribution of Wild Species to the Domesticated Apple Maternal Line." *Molecular Biology and Evolution* 30, no. 8 (August): 1751–1760.

Torgrimson, John. 2009. *Fruit, Berry, and Nut Inventory*. 4th ed. Seed Savers Exchange.

16. True Wild

Antonelli, Alexandre. 2022. *The Hidden Universe: Adventures in Biodiversity*. University of Chicago Press.

Dzhangeliev, A. D. 2002. "The Wild Apple Tree of Kazakhstan." In *Horticultural Reviews: Wild Apple and Fruit Trees of Central Asia* 29. Edited by Jules Janick. Wiley.

Torgrimson, John. 2009. *Fruit, Berry, and Nut Inventory*. 4th ed. Seed Savers Exchange.

17. Love Notes from Nature

The American Southwest. n.d. "Fallugia Paradoxa, Apache Plume. https://www.americansouthwest.net/plants/wildflowers/fallugia-paradoxa.html.

DeGraaf, Richard M., and Gretchin M. Witman. 1979. *Trees, Shrubs, and Vines for Attracting Birds*. University of Massachusetts Press.

Gibbons, Euell. 2020. *Stalking the Wild Asparagus*. Stackpole Books.

Wilson, E. O. 1993. *The Diversity of Life*. W. W. Norton.

18. Summer Lake

Daumal, René. 2023. *Rasa or Knowledge of the Self: Essays on Indian Aesthetics and Selected Sanskrit Studies*. Translated by Louise Landes Levi. Cool Grove Press.

Foster, Teressa, 1998. *Settlers in Summer Lake Valley*. Maverick Publications.

Torgrimson, John. 2009. *Fruit, Berry, and Nut Inventory*. 4th ed. Seed Savers Exchange.

19. The Hidden Rose

d'Aulaire, Ingri, and Edgar Parin d'Aulaire. 1992. *D'Aulaires' Book of Greek Myths*. Delacorte Press.

Littlewood, A. R. 1968. "The Greek Philosopher: The Symbolism of the Apple in Greek and Roman Literature." *Harvard Studies in Classical Philology* 72:147–181.

Longfellow, Henry Wadsworth., trans. 1867. *Dante Alighieri: Inferno*. Penguin Random House.

Torgrimson, John. 2009. *Fruit, Berry, and Nut Inventory*. 4th ed. Seed Savers Exchange.

20. The Songs of Ancient Trees

Ivey, James E. 1988. *In the Midst of a Loneliness: The Architectural History of the Salinas Missions*. Southwest Cultural Resources Center.

McLemore, Virginia. 2000. "Manzano Mountains State Park and the Abó and Quarai Units of the Salinas Pueblo Missions National Monument." *New Mexico Geology* 22:108–112.

Morrow, Baker H., ed. and trans. 2012. *A Harvest of Reluctant Souls: Fray Alonso de Benavides's History of New Mexico, 1630*. University of New Mexico Press.

21. Newtown

Albermarle Ciderworks. n.d. "Albemarle Pippin." https://www.albemarleciderworks.com/orchard/apple/albemarle-pippin.

Bear, James, and Lucia Stanton, eds. 2017. *The Papers of Thomas Jefferson*: *Jefferson's Memorandum Books, Accounts, with Legal Records and Miscellany*. Second Series. Princeton University Press.

Betts, Edwin Morris, ed. 1999. *Thomas Jefferson's Garden Book*. University of North Carolina Press.

Correspondence from Thomas Jefferson to Thomas Mann Randolph, Jr., 27 August 1786. https://founders.archives.gov/documents/Jefferson/01-10-02-0226.

Dolbier, Alison. 2023. "The Jefferson Weather and Climate Records." The Papers of Thomas Jefferson. October 2. https://jeffersonpapers.princeton.edu/welcome/the-jefferson-weather-and-climate-records/.

Hatch, Peter J. 2007. *The Fruits and Fruit Trees of Monticello*. University of Virginia Press.

National Archives. n.d. "The Declaration of Independence: What Does It Say?" America's Founding Documents. https://www.archives.gov/founding-docs/declaration/what-does-it-say.

Raver, Anne. 2013. "Wanted Dead or Alive (No, Just Dead)." *The New York Times*, November 27.

Stanton, Lucia. 2012. *"Those Who Labor for My Happiness": Slavery at Thomas Jefferson's Monticello*. University of Virginia Press.

Torgrimson, John. 2009. *Fruit, Berry, and Nut Inventory*. 4th ed. Seed Savers Exchange.

About the Author

Priyanka Kumar is the author of *Conversations with Birds*, widely acclaimed as "a landmark book" that "could help people around the world rewild their hearts and souls" (*Psychology Today*). Kumar's essays appear in *The New York Times*, *The Washington Post*, *Los Angeles Review of Books*, *Orion*, and *Sierra Magazine*. She has been featured on CBS News Radio, Yale Climate Connections, and Oprah Daily, and her awards include an Alfred P. Sloan Foundation Award, an ICJS Fellowship, a New Mexico/New Visions Governor's Award, an Aldo Leopold residency, a Canada Council for the Arts Grant, and an Academy of Motion Pictures Arts and Sciences Fellowship. She holds an MFA from the University of Southern California's School of Cinematic Arts and is an alumna of the Bread Loaf Writers' Conference. She wrote, directed, and produced the feature documentary *The Song of the Little Road*, starring Martin Scorsese and Ravi Shankar. Kumar taught at the University of California, Santa Cruz and the University of Southern California and serves on the Advisory Council of the Leopold Writing Program and on the Board of New Mexico Writers. She completed a Climate Master certification in 2025.